Springer Theses

Recognizing Outstanding Ph.D. Research

For further volumes:
http://www.springer.com/series/8790

Aims and Scope

The series "Springer Theses" brings together a selection of the very best Ph.D. theses from around the world and across the physical sciences. Nominated and endorsed by two recognized specialists, each published volume has been selected for its scientific excellence and the high impact of its contents for the pertinent field of research. For greater accessibility to non-specialists, the published versions include an extended introduction, as well as a foreword by the student's supervisor explaining the special relevance of the work for the field. As a whole, the series will provide a valuable resource both for newcomers to the research fields described, and for other scientists seeking detailed background information on special questions. Finally, it provides an accredited documentation of the valuable contributions made by today's younger generation of scientists.

Theses are accepted into the series by invited nomination only and must fulfill all of the following criteria

- They must be written in good English.
- The topic should fall within the confines of Chemistry, Physics, Earth Sciences, Engineering and related interdisciplinary fields such as Materials, Nanoscience, Chemical Engineering, Complex Systems and Biophysics.
- The work reported in the thesis must represent a significant scientific advance.
- If the thesis includes previously published material, permission to reproduce this must be gained from the respective copyright holder.
- They must have been examined and passed during the 12 months prior to nomination.
- Each thesis should include a foreword by the supervisor outlining the significance of its content.
- The theses should have a clearly defined structure including an introduction accessible to scientists not expert in that particular field.

Marcin Mucha-Kruczyński

Theory of Bilayer Graphene Spectroscopy

Doctoral Thesis accepted by
the University of Lancaster, UK

 Springer

Author
Dr. Marcin Mucha-Kruczyński
University of Lancaster
Lancaster
UK

Supervisor
Prof. Vladimir I. Falko
University of Lancaster
Lancaster
UK

ISSN 2190-5053   ISSN 2190-5061   (electronic)
ISBN 978-3-642-44673-3   ISBN 978-3-642-30936-6 (eBook)
DOI 10.1007/978-3-642-30936-6
Springer Heidelberg New York Dordrecht London

Printed on acid-free paper

Springer is part of Springer Science+Business Media (www.springer.com)

Supervisor's Foreword

Marcin Mucha-Kruczyński's Ph.D. Thesis presents the theory of three key elements of optical spectroscopy of the electronic excitations in a new material called bilayer graphene: angle-resolved photoemission spectroscopy (ARPES), electronic contribution to the visible range Raman spectroscopy (ERS), and far-infrared (FIR) magneto-spectroscopy. Bilayer graphene consists of two coupled honeycomb layers of carbon atoms and is a close relative of graphene, single layer of carbons. In the short time span of several years since their discovery, graphene materials have shown themselves to be extraordinary—they combine reduced dimensionality with the properties of the carbon–carbon sp^2 bond, to produce the thinnest yet strongest materials found so far, with high electron mobility and thermal conductivity. One of the peculiarities of graphene is that the electrons in the crystal behave like chiral particles. This chirality can be characterized by the topological Berry phase of electrons, which, in turn, depends on the number of graphene layers in the material. The most striking result presented in this Thesis is the prediction that this electronic chirality can be 'photographed' with ARPES, both for monolayer and bilayer crystals.

It turns out there are reasons to single bilayer graphene out of the group of graphene materials and to devote a whole Ph.D. Thesis to it. The coupling of two honeycomb layers of carbons placed on top of each other in the Bernal stacking, arrangement known from the study of graphite, is already enough to significantly modify the electronic dispersion in the range of energies relevant for transport as compared to the single layer graphene. Even more importantly, many of the properties of bilayer graphene can be controlled by an externally applied transverse electric field which shifts part of the charge carriers between the two carbon layers and opens a gap in the electronic spectrum. This feature is unique to bilayers and needs to be taken into account whenever top/bottom gates are used to modulate the density of carriers in the material. Marcin has shown how such a gap would manifest itself in ARPES measurements. He also investigated the FIR spectra of bilayer graphene placed in a strong external magnetic field applied perpendicularly to the carbon layers and kept at a set filling factor with the help of gates. To do so, he studied in detail Landau-level structure in bilayer graphene and constructed a scheme to self-consistently calculate the charge asymmetry in magnetic field in order to simulate the experimental situation. It is worth emphasizing that all of these results were contrasted to and confirmed by experiments.

Finally, he also investigated the electronic contribution to the Raman spectra of bilayer graphene. Raman spectroscopy is widely used to characterize carbon materials and in case of graphene careful analysis of the spectrum yields, among others, information about the number of layers, domain sizes, and doping levels in the sample. Nevertheless, the purely electronic in origin contribution to the Raman spectrum has been, initially, ignored. Marcin offers a comprehensive theory of inelastic Raman scattering resulting in the electron–hole excitations in bilayer graphene, at zero and quantizing magnetic fields. He predicts polarization properties of the dominant excitations and determines peculiar selection rules for the leading modes in the magnetic field in terms of the inter-Landaulevel transitions as well as their intensity. All those predictions have been confirmed by experiments performed after the completion of the Thesis [1, 2].

This Thesis should be useful to those interested in the properties of bilayer graphene and graphene optics.

Lancaster, UK, April 2012   Prof. V. I. Falko

References

1. C. Faugeras et al., Magneto-Raman scattering of graphene on graphite: electronic and phonon excitations. Phys. Rev. Lett. **107**, 036807 (2011)
2. P. Kossacki et al., Electronic excitations and electron-phonon coupling in bulk graphite through Raman scattering in high magnetic fields. Phys. Rev. B **84**, 235138 (2011)

Acknowledgments

The story of my Ph.D. starts much before the official admission date to Lancaster University. However, through all of it, two people played dominant roles. The first is Prof. V. I. Falko, who six years ago gave series of lectures during a summer school in a quite remote city of Jyväskylä. He later helped one of the students who attended that course to study at Lancaster University for a year as an exchange student. Three and a half years ago the student came back to Lancaster to start his Ph.D. under Prof. Falko's guidance. Without this guidance and his always-present fascination with physics, I would not have arrived to where I am now. The second person I would like to thank is my joint supervisor Dr. Edward McCann, who more than once discussed and helped me understand those 'simple and obvious' details I was stuck at, whether as an undergrad or as a finishing Ph.D. student. Acknowledgements are also due to O. Kashuba, Y. Suprunenko, and all those unnamed here, who through everyday discussions contributed toward my learning. I extend special thanks to my family, who always unanimously supported me in my decision to come to Lancaster. Last but not least, a special thank you to my wife Amie, who at times was probably the most interested of all involved in the finalization of this thesis.

Contents

1 Introduction 1
  1.1 Graphene on Paper and in the Lab. 1
  1.2 Two Layers: Double the Fun?. 2
  1.3 Thesis Outline. 3
  References 5

2 The Tight-Binding Approach and the Resulting
Electronic Structure. 9
  2.1 The Crystal and Reciprocal Lattices. 9
  2.2 The Four-Band Tight-Binding Model for π Electrons. 10
     2.2.1 Full Momentum Dependence 10
     2.2.2 Approximation for Hopping Elements 14
     2.2.3 Symmetry-Breaking Asymmetries in the
           on-Site Energies 16
  2.3 The Effective Two-Band Model 18
  References 20

3 Angle-Resolved Photoemission Spectroscopy 23
  3.1 ARPES as Quantum Young's Experiment 24
  3.2 Monolayer Graphene 27
  3.3 Bilayer Graphene. 29
     3.3.1 Low-Energy Spectrum: Contribution of the
           Two Degenerate Bands Only 30
     3.3.2 Contribution from the Split Bands. 31
     3.3.3 Influence of the Symmetry-Breaking Parameters
           on the ARPES Spectra. 34
     3.3.4 Interference Due to a Finite Interlayer Distance 35
  References 36

4 Magneto-Optical Spectroscopy 39
    4.1   Bilayer Graphene in an External Magnetic Field 41
        4.1.1   Landau Levels in the Two-Band Model 41
        4.1.2   Landau Levels in the Four-Band Model 42
        4.1.3   Numerical Treatment of the γ_3 Coupling 47
    4.2   Magneto-Optical Selection Rules and the Absorption Spectra ... 49
    4.3   Magneto-Optical Spectroscopy in Charged Bilayer Graphene ... 52
        4.3.1   Landau Level Spectrum in Charged Bilayer Graphene:
               Self-Consistent Analysis of the Interlayer
               Asymmetry Gap 52
        4.3.2   Tracking a Single Inter-LL Transition: Low-Energy
               Inter-Landau Level Transitions 56
        4.3.3   Magneto-Optical Spectra in Charged Bilayer:
               High-Energy Inter-Landau Level Transitions 58
    References 60

5 Electronic Raman Spectroscopy 63
    5.1   General Considerations 64
    5.2   The Two-Photon Field and the Electron-Photon Interaction 65
    5.3   Scattering Amplitude of the ERS Process 67
        5.3.1   Contribution of the Contact Interaction 68
        5.3.2   Contribution of the Two-Step Processes 68
        5.3.3   The Final Form of the Raman Scattering Amplitude 69
    5.4   ERS Spectra in the Absence of the Magnetic Field 70
    5.5   ERS Spectra in Quantizing Magnetic Fields 71
    References 74

6 Conclusions 77
    References 78

Curriculum Vitae 81

Chapter 1
Introduction

1.1 Graphene on Paper and in the Lab

With many models in physics, it is much easier to conceive a Gedankenexperiment and analyse it on paper, rather than prepare a real-life experiment. It is definitely the case when imagining a single plane of carbon atoms arranged in a honeycomb (hexagonal) pattern. As a conceptual building block of graphite [1], this model has been used by theorists to explain this material's physical and chemical properties for more than sixty years [2–4]. It resurfaced now and again, especially with the discovery of fullerenes [5] and tremendous interest following the rediscovery of carbon nanotubes [6, 7], both of which can be thought of as constructed from a layer of tightly arranged benzene rings. Somewhere along the way, the model got a name—graphene, signalling the presence of the planar sp^2 bonds between carbon atoms and emphasizing its importance in connection to graphite.

In the end, bulk graphite that was the reason for the Gedankenexperiment, had the main role in the real-life one. In 2004, Andre Geim's group at the University of Manchester, experimenting at the time with mechanical exfolation of layers from layered materials, isolated few-layer graphene films, including a single layer, from thin samples of highly-oriented pyrolytic graphite [8, 9]. Sheets of carbon, one atom thick, have been shown to be stable under ambient environment and were successfully processed into devices allowing for the investigation of their transport properties. These were found promising for both fundamental and application-oriented research: they confirmed theoretically predicted gapless linear dispersion of the quasiparticles in the vicinity of the Fermi energy, electron mobility $\sim 10^4 \frac{\mathrm{cm}^2}{\mathrm{Vs}}$, the mean free path of the order of tenths of micrometer, huge sustainable currents $> 10^8 \frac{\mathrm{A}}{\mathrm{cm}^2}$ and unusual sequence of plateaus in the Quantum Hall Effect associated with an additional electronic degree of freedom due to the symmetry of the crystal lattice [8, 10, 11]. Those first experiments attracted, therefore, a huge interest in the condensed matter community. In six years, their seminal Science paper, Ref. [8],

M. Mucha-Kruczyński, *Theory of Bilayer Graphene Spectroscopy*,
Springer Theses, DOI: 10.1007/978-3-642-30936-6_1,
© Springer-Verlag Berlin Heidelberg 2013

has been cited more than 2,800 times.[1] The on-line archive arXiv.org alone lists
over two thousand papers on widely defined graphene systems (monolayer, bilayer,
few-layer films, nanoribbons, etc.) submitted since 2007. Multiple general [12–16]
and more detailed reviews on synthesis [17, 18], optical [19], Raman [20, 21] and
photoemission [22] studies and electronic structure and transport [23–28] are already
available in the literature.

Other than mechanical exfoliation, ways to obtain graphene were explored and
advances in the epitaxial growth of carbon crystallites and layers on SiC and metallic
substrates [29–31] were taken advantage of. Note that although in some of those cases
prior to 2004 single layers of carbon atoms have been grown on a substrate, their
properties (for example the electronic dispersion at the K point) were dissimilar to
monolayer graphene. Currently, after significant development, graphene-like layers
can also be grown epitaxially on multiple substrates like SiC [32–34], Ni [35–38],
Ir [39], Ru [40–42] or Cu [43]. On some of these, only monolayer or effectively
monolayer-like decoupled layers can be grown while others allow for growth of few-
layer graphene films in various stackings. The technology is already advanced enough
to produce coverage areas of the order of square inch [43–48]. Reliable and detailed
tools for characterisation of all graphene-like systems are necessary to evaluate their
quality quickly and efficiently. At the same time, extensive knowledge of and about
the materials is needed to design future devices and engineer new graphene-based
systems. This thesis describes efforts undertaken to define a minimal theoretical
model, allowing for such prediction, for one of the many materials from the graphene
family, that is bilayer graphene.

1.2 Two Layers: Double the Fun?

Bilayer graphene was first obtained with the scotch-tape technique, originally used
to obtain monolayer graphene. A very 'messy' method, it yields at the final stage all
kinds of graphene-based thin films with varying thicknesses. Some of them consist of
two coupled layers of graphene. The most energetically favourable relative arrangement of those layers is the AB (or Bernal) stacking [49], also found in crystalline
graphite [1, 50]. McCann and Fal'ko showed that the low-energy electronic spectrum
of such a system, relevant for transport experiments, is qualitatively different from
the monolayer [51]. Most significantly, the interlayer coupling changes the dispersion from linear to quadratic. The resulting quasiparticles now exhibit Berry phase
of 2π leading to, for example, weak localisation instead of weak anti-localisation as
in the monolayer [52, 53]. However, at least in the neutral system, the spectrum is
still gapless. This last fact, together with the quadratic spectrum leads to the presence of an eightfold, including spin, degenerate Landau level (LL) positioned at zero
energy (Fermi energy in the neutral structure) and yet another unusual sequencing
of the plateaus in the Quantum Hall Effect as the plateau at $\sigma_{xy} = 0$ is missing [54].

---

[1] According to Thomson's ISI Web of Knowledge data.

McCann and Fal'ko were also the first to point out the possibility of breaking the layer symmetry by applying an external electric field. This gap was first directly observed with the help of the angle-resolved photoemission by Ohta et al. [55]. Thus, bilayer graphene became technologically relevant as one of the several options of introducing a gap in the electronic spectrum of graphene-based systems. On a more basic level, bilayer graphene introduced to solid state physics the unique notion of massive chiral fermions [15].

1.3 Thesis Outline

As explained in the previous section, bilayer graphene, despite its name and origins, can definitely be considered a material on its own rights, rather than a poor cousin of (monolayer) graphene. As the experimental effort in its characterisation grows and first devices are built, it is desirable to construct a theoretical model, simple yet universal, capturing essential physics in the wide scope of laboratory-relevant situations. In this thesis, we probe the limits and capabilities of the tight-binding model constructed for the π electrons only. We use this approximation to describe theoretically results of selected spectroscopic investigations of bilayer graphene, namely the angle-resolved photoemission spectroscopy (ARPES), magneto-optical spectroscopy (MOS) and electronic Raman spectroscopy (ERS). In particular, we concentrate on the comparison of the so called four-band and two-band models as we search for the 'minimal model' description. We also investigate the importance of the layer and sublattice symmetry breaking terms.

We start with the introduction of the tight-binding model of π electrons for graphene systems, presented in Chap. 2. We discuss in detail, the four-band model describing the π bands within the whole Brillouin zone and the linear approximation valid only in the vicinity of its corners and define the symmetry-breaking parameters contained within the model. We also describe the effective two-band, low-energy approximation for the band structure of bilayer graphene, which is the start point for a significant part of the theoretical models postulated in the literature to describe the physics of bilayer graphene.

The choice of spectroscopies for which we aim to provide theoretical description of corresponding spectra, has been mainly dictated by the developments in the experimental characterisation of graphene materials. The angle-resolved photoemission studies have been performed extensively on epitaxially grown monolayer graphene (e.g., [34, 35, 39, 56, 57]). Photoemission is one of the methods used, for example, to confirm the linearity and lack/presence of gaps in the electronic spectrum of graphene-like materials. The ARPES spectra of bilayer graphene have been used to demonstrate directly the presence of the gap in the electronic spectrum as the interlayer asymmetry has been introduced [55]. Our model [58], presented in Chap. 3, examines the angular distribution of the constant-energy maps of the ARPES intensity for monolayer and bilayer graphene. We show how these are related to the chirality of electrons in those systems. Afterwards, we show that for bilayer graphene, electronic states belonging to the high-energy bands impact the intensity of the spectra even when the experiment probes range of energies (as measured from the neutrality point) usually considered to be well described by the two-band approximation. Furthermore, we explain how the anisotropy of the constant-energy maps may be used to extract information about the magnitude and sign of interlayer coupling parameters and about symmetry breaking inflicted on the bilayer by the underlying substrate.

The second spectroscopic method considered in this thesis, magneto-optical absorption spectroscopy, is one of the experimental tools employed to examine the electronic states in graphene systems in external magnetic fields [59–68]. In Chap. 4, we investigate some of the aspects of magneto-optical spectra from the theoretical point of view. We start with the description of the Landau level structure in bilayer graphene. We complete the theory based on the two-band model [69, 70] and give selection rules as well as the optical strengths of the inter-Landau-level excitations taking into account all four π bands and the physically most relevant asymmetries [71]. We then look closer at the experimental setup used to probe the Landau level structure in bilayer graphene and discuss the importance of the external gates used to vary the carrier density in the bilayer during the experiment. Building on the theory proposed for the case of no magnetic field [72], we then present a self-consistent calculation of the electric-field-induced interlayer asymmetry in magnetic fields and the resulting Landau level structure. We also analyse the magneto-optical spectra of bilayer flakes in the photon-energy range corresponding to transitions between degenerate and split bands of bilayers [73].

Finally, in Chap. 5, we turn towards Raman spectroscopy. Routinely used to characterise carbon materials, in graphene systems in particular, it provides information on, for example, the number of layers, domain sizes, doping levels, thermal conductivity and the structure of edges [20, 21]. Here, we concentrate on relatively unexplored in graphene materials, purely electronic in origin, processes leading to inelastic scattering of light from the sample. The experimental approach focusing on such processes is often called electron Raman scattering/spectroscopy (ERS) [74]. We study the contribution of the low-energy electronic excitations toward the Raman spectrum of bilayer graphene for the incoming photon energy $\Omega \gg 1\,\mathrm{eV}$ both in the presence and absence of an external magnetic field [75]. Starting with the four-band tight-binding model, we derive an effective scattering amplitude that can be incorporated into the two-band approximation. We show that this effective scattering amplitude is different from the contact interaction amplitude obtained within the two-band model alone. We then calculate the spectral density of the inelastic light scattering accompanied by the excitation of electron-hole pairs in bilayer graphene. In the absence of a magnetic field, due to the parabolic dispersion of the low-energy bands in a bilayer crystal, this contribution is constant and in doped structures has a threshold at twice the Fermi energy. In an external magnetic field, the dominant Raman-active modes are the $n_- \rightarrow n_+$ inter-Landau-level transitions with crossed polarization of in/out photons. We estimate the quantum efficiency of a single $n_- \rightarrow n_+$ transition in the magnetic field of $10\,\mathrm{T}$ as $I_{n_- \rightarrow n_+} \sim 10^{-12}$, which may be experimentally observable.

We summarise the work presented in this thesis in Chap. 6. Based on Chaps. 3, 4 and 5, we discuss the applicability of the two- and four-band models with respect to the electronic structure of bilayer graphene around the Fermi energy and prediction of experimental measurements.

References

1. J.D. Bernal, The structure of graphite. Proc. Royal Soc. A **106**, 749 (1924)
2. P.R. Wallace, The band theory of graphite. Phys. Rev. **71**, 622 (1947)
3. J.W. McClure, Band structure of graphite and de Haas-van Alphen effect. Phys. Rev. **108**, 612 (1957)
4. J.C. Slonczewski, P.R. Weiss, Band structure of graphite. Phys. Rev. **109**, 272 (1958)
5. H.W. Kroto, J.R. Heath, S.C. O'Brien, R.F. Curl, R.E. Smalley, C_{60}: Buckminsterfullerene. Nature **318**, 162 (1985)
6. S. Ijima, Helical microtubules of graphite carbon. Nature **354**, 56 (1991)
7. S. Ijima, T. Ichihashi, Single-shell carbon nanotubes of 1-nm diameter. Nature **363**, 603 (1993)
8. K.S. Novoselov, A.K. Geim, S.V. Morozov, D. Jiang, Y. Zhang, S.V. Dubonos, I.V. Grigorieva, A.A. Firsov, Electric field effect in atomically thin carbon films. Science **306**, 666 (2004)
9. K.S. Novoselov, D. Jiang, T. Booth, V.V. Khotkevich, S.V. Morozov, A.K. Geim, Two-dimensional atomic crystals. Proc. Natl Acad. Sci. U S A **102**, 10451 (2005)
10. K.S. Novoselov, A.K. Geim, S.V. Morozov, D. Jiang, M.I. Katsnelson, I.V. Grigorieva, S.V. Dubonos, A.A. Firsov, Two-dimensional gas of massless dirac fermions in graphene. Nature **438**, 197 (2005)
11. Y. Zhang, Y.W. Tan, H.L. Stormer, P. Kim, Experimental observation of quantum hall effect and Berry's phase in graphene. Nature **438**, 201 (2005)
12. M.I. Katsnelson, Graphene: carbon in two dimensions. Mater. Today **10**, 20 (2007)
13. A.K. Geim, K.S. Novoselov, The rise of graphene. Nat. Mater. **6**, 183 (2007)
14. M.I. Katsnelson, K.S. Novoselov, Graphene: New bridge between condensed matter physics and quantum electrodynamics. Solid State Commun. **143**, 3 (2007)
15. A.K. Geim, Graphene: Status and prospects. Science **324**, 1530 (2009)
16. D.S.L. Abergel, V. Apalkov, J. Berashevich, K. Ziegler, T. Chakraborty, Properties of graphene: a theoretical perspective. Adv. Phys. **59**, 261 (2010)
17. J. Hass, W.A. de Heer, E.H. Conrad, The growth and morphology of epitaxial multilayer graphene. J. Phys. Condens. Matter **20**, 323202 (2008)
18. J. Wintterlin, M.-L. Bocquet, Graphene on metal surfaces. Surf. Sci. **603**, 1841 (2009)
19. M. Orlita, M. Potemski, Dirac electronic states in graphene systems: optical spectroscopy studies. Semicond. Sci. Technol. **25**, 063001 (2010)
20. A.C. Ferrari, Raman spectroscopy of graphene and graphite: Disorder, electron phonon coupling, doping and nonadiabatic effects. Solid State Commun. **143**, 47 (2007)
21. N. Ferralis, Probing mechanical properties of graphene with Raman spectroscopy. J. Mater. Sci. **97**, 5135 (2010)
22. A. Bostwick, K.V. Emtsev, K. Horn, E. Huwald, L. Ley, J.L. McChesney, T. Ohta, J. Riley, E. Rotenberg, F. Speck, T. Seyller, Photoemission studies of graphene on SiC: growth, interface, and electronic structure. Adv. Solid State Phys. **47**, 159 (2008)
23. K.S. Novoselov, S.V. Morozov, T.M.G. Mohinddin, L.A. Ponomarenko, D.C. Elias, R. Yang, I.I. Barbolina, P. Blake, T.J. Booth, D. Jiang, J. Giesbers, E.W. Hill, A.K. Geim, Electronic properties of graphene. Phys. Status Solibi B **244**, 4106 (2007)
24. C.W.J. Beenakker, Andreev reflection and Klein tunneling in graphene. Rev. Mod. Phys. **80**, 1337 (2008)
25. A.H. Castro Neto, F. Guinea, N.M.R. Peres, K.S. Novoselov, A.K. Geim, The electronic properties of graphene. Rev. Mod. Phys. **81**, 109 (2009)

26. A. Bostwick, J.L. McChesney, T. Ohta, E. Rotenberg, T. Seyller, K. Horn, Experimental studies of the electronic structure of graphene. Prog. Surf. Sci. **84**, 380 (2009)
27. N.M.R. Peres, The transport properties of graphene: an introduction. Rev. Mod. Phys. **82**, 2673 (2010)
28. S. Das Sarma, S. Adam, E.H. Hwang, E. Rossi, Electronic transport in two dimensional graphene, arXiv:1003.4731 (2010)
29. C. Oshima, A. Nagashima, Ultra-thin epitaxial films of graphite and hexagonal boron nitride on solid surfaces. J. Phys. Condens. Matter **9**, 1 (1997)
30. A. Charrier, A. Coati, T. Argunova, F. Thibaudau, Y. Garreau, R. Pinchaux, I. Forbeaux, J.-M. Debever, M. Sauvage-Simkin, J.-M. Themlin, Solid-state decomposition of silicon carbide for growing ultra-thin heteroepitaxial graphite films. J. Appl. Phys. **92**, 2479 (2002)
31. C. Berger, Z. Song, T. Li, X. Li, A.Y. Ogbazghi, R. Feng, Z. Dai, A.N. Marchenkov, E.H. Conrad, P.N. First, W.A. de Heer, Ultrathin epitaxial graphite: 2d electron gas properties and a route toward graphene-based nanoelectronics. J. Phys. Chem. B **108**, 19912 (2004)
32. C. Berger, Z. Song, X. Li, X. Wu, N. Brown, C. Naud, D. Mayou, T. Li, J. Hass, A.N. Marchenkov, E.H. Conrad, P.N. First, W.A. de Heer, Electronic confinement and coherence in patterned epitaxial graphene. Science **312**, 1191 (2006)
33. E. Rollings, G.-H. Gweon, S.Y. Zhou, B.S. Mun, J.L. McChesney, B.S. Hussain, A.V. Federov, P.N. First, W.A. de Heer, A. Lanzara, Synthesis and characterization of atomically thin graphite films on a silicon carbide substrate. J. Phys. Chem. Solids **67**, 2172 (2006)
34. M. Sprinkle, D. Siegel, Y. Hu, J. Hicks, A. Tejeda, A. Taleb-Ibrahimi, P. Le Fèvre, F. Bertran, S. Vizzini, H. Enriquez, S. Chiang, P. Soukiassian, C. Berger, W.A. de Heer, A. Lanzara, E.H. Conrad, First direct observation of a nearly ideal graphene band structure. Phys. Rev. Lett. **103**, 226803 (2009)
35. Y.S. Dedkov, M. Fonin, C. Laubschat, A possible source of spin-polarized electrons: the inert graphene/Ni(111) system. Appl. Phys. Lett. **92**, 052506 (2008)
36. Y.S. Dedkov, M. Fonin, U. Rüdiger, C. Laubschat, Rashba effect in the graphene/Ni(111) system. Phys. Rev. Lett. **100**, 107602 (2008)
37. A. Grüneis, D.V. Vyalikh, Tunable hybridization between electronic states of graphene and a metal surface. Phys. Rev. B **77**, 193401 (2008)
38. A. Grüneis, K. Kummer, D.V. Vyalikh, Dynamics of graphene growth on a metal surface: a time-dependent photoemission study. New J. Phys. **11**, 073050 (2009)
39. I. Pletikosić, M. Kralj, P. Pervan, R. Brako, J. Coraux, A.T. N'Diaye, C. Busse, T. Michely, Dirac cones and minigaps for graphene on Ir(111). Phys. Rev. Lett. **102**, 056808 (2009)
40. A.L. Vázquez de Parga, F. Calleja, B. Borca, M.C.G. Passegi Jr, J.J. Hinarejos, F. Guinea, R. Miranda, Periodically rippled graphene: growth and spatially resolved electronic structure. Phys. Rev. Lett. **100**, 056807 (2008)
41. P.W. Sutter, J.-I. Flege, E.A. Sutter, Epitaxial graphene on ruthenium. Nat. Mater. **7**, 406 (2008)
42. C. Enderlein, Y.S. Kim, A. Bostwick, E. Rotenberg, K. Horn, The formation of an energy gap in graphene on ruthenium by controlling the interface. New J. Phys. **12**, 033014 (2010)
43. Y. Lee, S. Bae, H. Jang, S. Jang, S.-E. Zhu, S.H. Sim, Y.I. Song, B.H. Hong, J.-H. Ahn, Wafer-scale synthesis and transfer of graphene films. Nano Lett. **10**, 490 (2010)
44. K.S. Kim, Y. Zhao, H. Jang, S.Y. Lee, J.M. Kim, K.S. Kim, J.-H. Ahn, P. Kim, J.-Y. Choi, B.H. Hong, Large-scale pattern growth of graphene films for stretchable transparent electrodes. Nature **457**, 706 (2009)
45. A. Reina, S. Thiele, X. Jia, S. Bhaviripudi, M.S. Dresselhaus, J.A. Schaefer, J. Kong, Growth of large-area single- and bi-layer graphene by controlled carbon precipitation on polycrystalline ni surfaces. Nano Res. **2**, 509 (2009)
46. H.J. Park, J. Meyer, S. Roth, V. Skakalova, Growth and properties of few-layer graphene prepared by chemical vapor deposition. Carbon **48**, 1088 (2010)
47. Z.-Y. Juang, C.-Y. Wu, A.-Y. Lu, C.-Y. Su, K.-C. Leou, F.-R. Chen, C.-H. Tsai, Graphene synthesis by chemical vapor deposition and transfer by a roll-to-roll process. Carbon **48**, 3169 (2010)

48. M. Xu, D. Fujita, K. Sagisaka, E. Watanabe, N. Hanagata, Single-layer graphene nearly 100 % covering an entire substrate, arXiv:1006.5085 (2010)
49. J.-C. Charlier, J.-P. Michenaud, X. Gonze, First-principles study of the electronic properties of simple hexagonal graphite. Phys. Rev. B **46**, 4531 (1992)
50. M.S. Dresselhaus, G. Dresselhaus, Intercalation compounds of graphite. Adv. Phys. **30**, 139 (1981)
51. E. McCann, V.I. Fal'ko, Landau level degeneracy and quantum hall effect in a graphite bilayer. Phys. Rev. Lett. **96**, 086805 (2006)
52. K. Kechedzhi, E. McCann, V. Fal'ko, B. Altshuler, Influence of trigonal warping on interference effects in bilayer graphene. Phys. Rev. Lett. **98**, 176806 (2007)
53. K. Kechedzhi, E. McCann, V. Fal'ko, H. Suzuura, T. Ando, B. Altshuler, Weak localization in monolayer and bilayer graphene. Eur. Phys. J. Special Top. **148**, 39 (2007)
54. K.S. Novoselov, E. McCann, S.V. Morozov, V.I. Fal'ko, M.I. Katsnelson, U. Zeitler, D. Jiang, F. Schedin, A.K. Geim, Unconventional quantum hall effect and Berry's phase of 2pi in bilayer graphene. Nat. Phys. **2**, 177 (2006)
55. T. Ohta, A. Bostwick, T. Seyller, K. Horn, E. Rotenberg, Controlling the electronic structure of bilayer graphene. Science **313**, 951 (2006)
56. A. Bostwick, T. Ohta, T. Seyller, K. Horn, E. Rotenberg, Quasiparticle dynamics in graphene. Nat. Phys. **3**, 36 (2007)
57. S.Y. Zhou, G.-H. Gweon, A.V. Fedorov, P.N. First, W.A. de Heer, D.-H. Lee, F. Guinea, A.H. Castro Neto, A. Lanzara, Substrate-induced bandgap opening in epitaxial graphene. Nat. Mater. **6**, 770 (2007)
58. M. Mucha-Kruczyński, O. Tsyplyatyev, A. Grishin, E. McCann, V.I. Fal'ko, A. Bostwick, E. Rotenberg, Characterization of graphene through anisotropy of constant-energy maps in angle-resolved photoemission. Phys. Rev. B **77**, 195403 (2008)
59. M.L. Sadowski, G. Martinez, M. Potemski, C. Berger, W.A. de Heer, Landau level spectroscopy of ultrathin graphite layers. Phys. Rev. Lett. **97**, 266405 (2006)
60. Z. Jiang, E.A. Henriksen, L.C. Tung, Y.-J. Wang, M.E. Schwartz, M.Y. Han, P. Kim, H.L. Stormer, Infrared spectroscopy of Landau levels of graphene. Phys. Rev. Lett. **98**, 197403 (2007)
61. R.S. Deacon, K.-C. Chuang, R.J. Nicholas, K.S. Novoselov, A.K. Geim, Cyclotron resonance study of the electron and hole velocity in graphene monolayers. Phys. Rev. B **76**, 081406(R) (2007)
62. M.L. Sadowski, G. Martinez, M. Potemski, C. Berger, W.A. de Heer, Magnetospectroscopy of epitaxial few-layer graphene. Solid State Commun. **143**, 123 (2007)
63. E.A. Henriksen, Z. Jiang, L.-C. Tung, M.E. Schwartz, M. Takita, Y.-J. Wang, P. Kim, H.L. Stormer, Cyclotron resonance in bilayer graphene. Phys. Rev. Lett. **100**, 087403 (2008)
64. P. Płochocka, C. Faugeras, M. Orlita, M.L. Sadowski, G. Martinez, M. Potemski, M.O. Goerbig, J.-N. Fuchs, C. Berger, W.A. de Heer, High-energy limit of massless Dirac fermions in multilayer graphene using magneto-optical transmission spectroscopy. Phys. Rev. Lett. **100**, 087401 (2008)
65. M. Orlita, C. Faugeras, P. Plochocka, P. Neugebauer, G. Martinez, D.K. Maude, A.-L. Barra, M. Sprinkle, C. Berger, W.A. de Heer, M. Potemski, Approaching the Dirac point in high mobility multi-layer epitaxial graphene. Phys. Rev. Lett. **101**, 267601 (2008)
66. P. Neugebauer, M. Orlita, C. Faugeras, A.-L. Barra, M. Potemski, How perfect can graphene be? Phys. Rev. Lett. **103**, 136403 (2009)
67. E.A. Henriksen, P. Cadden-Zimansky, Z. Jiang, Z.Q. Li, L.-C. Tung, M.E. Schwartz, M. Takita, Y.-J. Wang, P. Kim, H.L. Stormer, Interaction-induced shift of the cyclotron resonance of graphene using infrared spectroscopy. Phys. Rev. Lett. **104**, 067404 (2010)
68. A.M. Witowski, M. Orlita, R. Stepniewski, A. Wysmolek, J.M. Baranowski, W. Strupinski, C. Faugeras, G. Martinez, M. Potemski, Quasi-classical cyclotron resonance of Dirac fermions in highly doped graphene, arXiv:1007.4153 (2010)
69. D.S.L. Abergel, V.I. Fal'ko, Optical and magneto-optical far-infrared properties of bilayer graphene. Phys. Rev. B **75**, 155430 (2007)

70. D.S.L. Abergel, E. McCann, V.I. Fal'ko, QHE and far infra-red properties of bilayer graphene in a strong magnetic field. Eur. Phys. J. Special Top. **148**, 105 (2007)
71. M. Mucha-Kruczyński, D.S.L. Abergel, E. McCann, V.I. Fal'ko, On spectral properties of bilayer graphene: the effect of an SiC substrate and infrared magneto-spectroscopy. J. Phys. Condens. Matter **21**, 344206 (2009)
72. E. McCann, Asymmetry gap in the electronic band structure of bilayer graphene, Phys. Rev. B **74**, 161403(R) (2006)
73. M. Mucha-Kruczyński, E. McCann, V.I. Fal'ko, Influence of interlayer asymmetry on magneto-spectroscopy of bilayer graphene. Solid State Commun. **149**, 1111 (2009)
74. M.V. Klein, Electronic Raman scattering, in *Light Scattering in Solids*, vol. I, ed. by M. Cardona (Springer, Berlin, 1983)
75. M. Mucha-Kruczyński, O. Kashuba, V.I. Fal'ko, Spectral features due to inter-Landau-level transitions in the Raman spectrum of bilayer graphene. Phys. Rev. B **82**, 045405 (2010)

Chapter 2
The Tight-Binding Approach and the Resulting Electronic Structure

In this chapter, we describe the crystal and reciprocal lattices of bilayer graphene. We also discuss briefly the symmetry of the crystal lattice. We then introduce the tight-binding model for π electrons in bilayer graphene. We start with a general formulation valid for all points in the Brillouin zone and the resulting electronic structure. Next, we concentrate on the linear approximation of that model around the corners of the Brillouin zone. This tight-binding approach is a variation of the tight-binding model for monolayer graphene as developed historically for applications in the physics of graphite (then so called Slonczewski–Weiss–McClure model [1–4]). For an introduction to the tight-binding approach in carbon sp^2 materials, see Refs. [5] or [6]. In the following section, we introduce symmetry-breaking parameters which will later turn out to be very important when interpreting results of spectroscopic measurements. We conclude the chapter with the derivation of the effective low-energy, two-band Hamiltonian for bilayer graphene.

2.1 The Crystal and Reciprocal Lattices

Bilayer graphene consists of two coupled graphene layers of carbon atoms (graphene monolayers) arranged in Bernal (AB) stacking [7, 8]. The unit cell contains four inequivalent carbon sites $A1$, $B1$, $A2$, and $B2$, where A and B denote two triangular sublattices in the same layer while 1 and 2 distinguish between the bottom and top layer, respectively. The real lattice of bilayer graphene is schematically shown in Fig. 2.1a. The honeycomb lattice of the bottom and top layers has been drawn with red and black solid lines, respectively. The lattice constant a, that is the $A - A$ (or $B - B$) distance, marked in the figure with grey, equals 2.46 Å. This lattice constant derives from the benzene-ring structure and is the same in bilayer graphene as in monolayer graphene or graphite. The interlayer distance, c_0, is much greater than the nearest neighbour carbon-carbon distance $\frac{a}{\sqrt{3}}$ Å. X-ray reflectivity experiments and first-principles calculations performed for bilayer graphene epitaxialy grown on

M. Mucha-Kruczyński, *Theory of Bilayer Graphene Spectroscopy,*
Springer Theses, DOI: 10.1007/978-3-642-30936-6_2,
© Springer-Verlag Berlin Heidelberg 2013

Table 2.1 Components of the vectors in the real and reciprocal lattices shown in Fig. 2.1 and used throughout the text

vector component	a_1	a_2	d_1	d_2	d_3	b_1	b_2	K_+	K_-
x or k_x	$\frac{a}{2}$	$\frac{a}{2}$	0	$\frac{a}{2}$	$-\frac{a}{2}$	$\frac{2\pi}{a}$	$\frac{2\pi}{a}$	$\frac{4\pi}{3a}$	$-\frac{4\pi}{3a}$
y or k_y	$\frac{a\sqrt{3}}{2}$	$-\frac{a\sqrt{3}}{2}$	$\frac{a}{\sqrt{3}}$	$-\frac{a}{2\sqrt{3}}$	$-\frac{a}{2\sqrt{3}}$	$\frac{2\pi}{a\sqrt{3}}$	$-\frac{2\pi}{a\sqrt{3}}$	0	0

SiC [9], as well as first-principles calculations for bilayer in vacuum [10], lead to $c_0 \approx 3.35\,\text{Å}$, as in graphite. We choose vectors a_1 and a_2 as our in-plane primitive lattice vectors and a rhombic unit cell as shown in Fig. 2.1a with a dashed blue line. Also shown in the figure are vectors d_1, d_2 and d_3, which can be used to express the distance between neighbouring in-plane carbon atoms. Eventually, we use vector $c_0 = (0, 0, c_0)$ to describe the thickness of the bilayer.

The corresponding reciprocal lattice is schematically presented in Fig. 2.1b. It is two-dimensional, because bilayer graphene, although strictly speaking three-dimensional due to the interlayer spacing, is not periodic in the z direction. The reciprocal unit vectors b_1 and b_2, related to a_1 and a_2 via the condition $b_i \cdot a_j = 2\pi \delta_{ij}$, are shown in darker blue. The Brillouin zone is a hexagon, marked in the figure with a dashed line. We denote two inequivalent corners of the Brillouin zone (later also called valleys) as K_+ (at the position $K_+ = (4\pi/3a, 0)$) and K_- (at the position $K_- = (-4\pi/3a, 0)$) and reserve index $\xi = \pm$ to distinguish between them in further discussions.

The real primitive lattice vectors a_1 and a_2, reciprocal primitive vectors b_1 and b_2, nearest neighbour vectors d_i, as well as the coordinates of the valley K_ξ are repeatedly used throughout the remaining parts of the thesis. For convenience, all aforementioned vectors and their components in their respective space have been summarised in Table 2.1.

2.2 The Four-Band Tight-Binding Model for π Electrons

2.2.1 Full Momentum Dependence

Let us consider an infinite sheet of bilayer graphene. For the origin of the coordinate system, we choose the centre of a unit cell (position of the $B1 - A2$ dimer, at the point halfway between the layers) and denote by r and R_0 the position vector and a vector pointing to the centre of another unit cell (one of N in total), respectively. We reserve symbol R_i to represent a vector pointing from the centre of a unit cell to the atomic site i (i then stands for $A1$, $B1$, $A2$ or $B2$) in this unit cell. We assume periodic boundary conditions and construct a basis of functions $\phi_{k,i}(r)$ built up from the π orbitals $\varphi(\mathbf{r})$ of carbon atoms in site i,

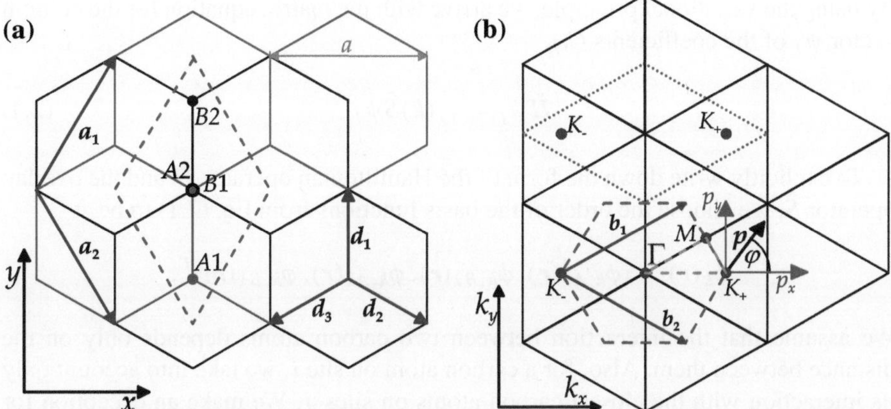

Fig. 2.1 a Schematic drawing of the bilayer graphene crystal lattice. The *bottom* (*top*) *layer* is depicted with *red* (*black*) *solid lines*. The real primitive lattice vectors are $\boldsymbol{a}_1$ and $\boldsymbol{a}_2$ and the unit cell is shown with *dashed blue line*. *Grey line* marks the lattice constant a. Vectors $\boldsymbol{d}_1$, $\boldsymbol{d}_2$, and $\boldsymbol{d}_3$, are used to express the relative position of neighbouring carbon atoms. **b** The reciprocal lattice of bilayer graphene with the Brillouin zone shown with the *dashed blue line* and its two inequivalent corners (valleys) K_+ and K_-. In contrast, the dotted blue line shows an alternative, rhombic unit cell in reciprocal space used briefly in Chap. 3. The *orange dashed line* shows the high-symmetry directions for which the band structure in Fig. 2.2 is shown

$$
\phi_{\boldsymbol{k},A1}(\boldsymbol{r}) = \frac{1}{\sqrt{N}} \sum_{\boldsymbol{R}_0} e^{i\boldsymbol{k}\cdot(\boldsymbol{R}_0 - \boldsymbol{d}_1 - \frac{c_0}{2})} \varphi\left(\boldsymbol{r} - \boldsymbol{R}_0 + \boldsymbol{d}_1 + \frac{c_0}{2}\right),
$$

$$
\phi_{\boldsymbol{k},B1}(\boldsymbol{r}) = \frac{1}{\sqrt{N}} \sum_{\boldsymbol{R}_0} e^{i\boldsymbol{k}\cdot(\boldsymbol{R}_0 - \frac{c_0}{2})} \varphi\left(\boldsymbol{r} - \boldsymbol{R}_0 + \frac{c_0}{2}\right),
$$

$$
\phi_{\boldsymbol{k},A2}(\boldsymbol{r}) = \frac{1}{\sqrt{N}} \sum_{\boldsymbol{R}_0} e^{i\boldsymbol{k}\cdot(\boldsymbol{R}_0 + \frac{c_0}{2})} \varphi\left(\boldsymbol{r} - \boldsymbol{R}_0 - \frac{c_0}{2}\right),
$$

$$
\phi_{\boldsymbol{k},B2}(\boldsymbol{r}) = \frac{1}{\sqrt{N}} \sum_{\boldsymbol{R}_0} e^{i\boldsymbol{k}\cdot(\boldsymbol{R}_0 + \boldsymbol{d}_1 + \frac{c_0}{2})} \varphi\left(\boldsymbol{r} - \boldsymbol{R}_0 - \boldsymbol{d}_1 - \frac{c_0}{2}\right),
\tag{2.1}
$$

where $\boldsymbol{k}$ is a two-dimensional electron wave vector.

The electron wave function $\Psi_j(\boldsymbol{r})$, corresponding to the energy eigenvalue $\epsilon_j(\boldsymbol{k})$ of an electron with wave vector $\boldsymbol{k}$, is a linear combination of functions in Eq. (2.1),

$$
\Psi_j(\boldsymbol{r}) = \sum_i C_{ij} \phi_{\boldsymbol{k},i}(\boldsymbol{r}).
\tag{2.2}
$$

It is easy to see that $\Psi_j(\boldsymbol{r})$ satisfies Bloch's theorem, as we have

$$
\Psi_j(\boldsymbol{r} + m\boldsymbol{a}_1 + n\boldsymbol{a}_2) = \sum_i C_{ij} \phi_{\boldsymbol{k},i}(\boldsymbol{r} + m\boldsymbol{a}_1 + n\boldsymbol{a}_2) = e^{i\boldsymbol{k}\cdot(m\boldsymbol{a}_1 + n\boldsymbol{a}_2)} \Psi_j(\boldsymbol{r}).
\tag{2.3}
$$

By using the variational principle, we arrive with the matrix equation for the column vector ψ_j of the coefficients C_{ij},

$$\hat{H}\psi_j = \epsilon_j(k)\hat{S}\psi_j. \tag{2.4}$$

To explicitly write down the form of the Hamiltonian operator $\hat{H}$ and the overlap operator $\hat{S}$, we choose the order of the basis functions from Eq. (2.1) to be

$$\phi_k(r) = (\phi_{k,A1}(r), \phi_{k,B2}(r), \phi_{k,A2}(r), \phi_{k,B1}(r))^T.$$

We assume that the interaction between two carbon atoms depends only on the distance between them. Also, for a carbon atom on site i, we take into account only its interaction with the closest carbon atoms on sites j. We make an exception for the interaction of an atom on site i with another on site i, where we include the influence of next-nearest neighbour of the same kind. The phase factor resulting from a summation over nearest neighbours can, for any carbon atom, be written in terms of the vectors d_1, d_2 and d_3. We define the geometrical factor $f(k)$,

$$f(k) \equiv \sum_{i=1}^{3} e^{ik\cdot d_i} = e^{i\frac{k_y a}{\sqrt{3}}} + 2e^{-i\frac{k_y a}{2\sqrt{3}}} \cos\frac{k_x a}{2}. \tag{2.5}$$

As a result, the full matrix form of the operators $\hat{H}$ and $\hat{S}$ is[1]:

$$\hat{H} = \begin{pmatrix} \epsilon_{A1} - \gamma_n|f(k)|^2 & -\gamma_3 f^*(k) & \gamma_4 f(k) & -\gamma_0 f(k) \\ -\gamma_3 f(k) & \epsilon_{B2} - \gamma_n|f(k)|^2 & -\gamma_0 f^*(k) & \gamma_4 f^*(k) \\ \gamma_4 f^*(k) & -\gamma_0 f(k) & \epsilon_{A2} - \gamma_n|f(k)|^2 & \gamma_1 \\ -\gamma_0 f^*(k) & \gamma_4 f(k) & \gamma_1 & \epsilon_{B1} - \gamma_n|f(k)|^2 \end{pmatrix}; \tag{2.6a}$$

$$\hat{S} = \begin{pmatrix} 1 & 0 & 0 & s_0 f(k) \\ 0 & 1 & s_0 f^*(k) & 0 \\ 0 & s_0 f(k) & 1 & s_1 \\ s_0 f^*(k) & 0 & s_1 & 1 \end{pmatrix}. \tag{2.6b}$$

In the above, we introduced several parameters into the model as a description of the strength of interactions between carbon atoms. In this, we mostly follow the Slonczewski-Weiss-McClure model developed for bulk graphite [1–4] (for a review see Ref. [8]). The on-site energies ϵ_i, couplings γ_j and overlap parameters s_l used in Eq. (2.6) are given by:

---

[1] We neglect in Eq. (2.6a) a factor of $3\gamma_n$ appearing on the diagonal as it only leads to a shift of zero on the energy scale.

$$\epsilon_i \equiv \langle \varphi(r - R_0 - R_i)|\hat{H}|\varphi(r - R_0 - R_i)\rangle, \tag{2.7a}$$

$$\gamma_0 \equiv -\langle \varphi(r - R_0 - R_{A1})|\hat{H}|\varphi(r - R_0 - R_{A1} - d_i)\rangle \tag{2.7b}$$

$$\equiv -\langle \varphi(r - R_0 - R_{A2})|\hat{H}|\varphi(r - R_0 - R_{A2} - d_i)\rangle,$$

$$\gamma_1 \equiv \langle \varphi(r - R_0 - R_{B1})|\hat{H}|\varphi(r - R_0 - R_{A2})\rangle, \tag{2.7c}$$

$$\gamma_3 \equiv -\langle \varphi(r - R_0 - R_{A1})|\hat{H}|\varphi(r - R_0 - R_{A1} + d_i - c_0)\rangle, \tag{2.7d}$$

$$\gamma_4 \equiv \langle \varphi(r - R_0 - R_{A1})|\hat{H}|\varphi(r - R_0 - R_{A1} - d_1 - c_0)\rangle \tag{2.7e}$$

$$\equiv \langle \varphi(r - R_0 - R_{B1})|\hat{H}|\varphi(r - R_0 - R_{B1} - d_1 - c_0)\rangle,$$

$$\gamma_n \equiv \langle \varphi(r - R_0 - R_i)|\hat{H}|\varphi(r - R_0 - R_i + d_3 - d_2)\rangle, \tag{2.7f}$$

$$s_0 \equiv \langle \varphi(r - R_0 - R_{A1})|\varphi(r - R_0 - R_{A1} - d_i)\rangle \tag{2.7g}$$

$$\equiv \langle \varphi(r - R_0 - R_{A2})|\varphi(r - R_0 - R_{A2} - d_i)\rangle,$$

$$s_1 \equiv \langle \varphi(r - R_0 - R_{B1})|\varphi(r - R_0 - R_{A2})\rangle. \tag{2.7h}$$

The diagonal terms ϵ_i denote the on-site energy of the electron at the carbon atom in site i. In the first approximation, they are equal to the energy of an electron in the $2p_z$ orbital of a carbon atom. This energy is modified as carbon atoms bond together to form the lattice. However, in an ideal and charge-neutral bilayer, this on-site energy is approximately the same for each site in the lattice. In this case, we can take it to be zero and define our energy scale relatively to this point. More complicated situations in which the symmetry between the atomic sites has been broken are discussed in Sect. 2.2.3.

The parameters γ_j describe the strength of the coupling between a specific pair of carbon atoms. The constant γ_0 denotes the coupling between the nearest neighbours ($A1 \leftrightarrow B1$ and $A2 \leftrightarrow B2$). The parameter γ_1 describes the direct interlayer coupling $A2 \leftrightarrow B1$. The γ_3 coupling represents the interlayer interaction between the nearest $A1$ and $B2$ atoms, whereas γ_4 characterizes the interlayer coupling between the nearest $A1$ and $A2$, as well as $B1$ and $B2$ atoms. Naively, one could expect the couplings γ_3 and γ_4 to be equal, especially with the assumptions made above about couplings dependent only on the interatomic distance. However, with some insight from the Slonczewski–Weiss–McClure model, we allow for γ_3 and γ_4 to be different. Physically, this is the case because, as opposed to γ_3, γ_4 involves one of the atomic sites creating the "dimer" ($B1$ and $A2$). The last coupling, γ_n, describes the interaction of the in-plane next-nearest neighbours.

The overlap integrals s_l take into account the fact that our π orbitals do not span an orthogonal basis set. We only included here the overlap s_0 between two nearest neighbour atoms and the overlap s_1 between the $A2$ and $B1$ sites where atoms are directly above/below each other. Due to their small value, in most of the situations under consideration in this thesis, even these two overlap integrals are neglected.

The electronic band structure resulting from Eq. (2.4) with the Hamiltonian and overlap matrices $\hat{H}$ and $\hat{S}$ as in Eq. (2.6) is shown in Fig. 2.2. We see two conduction and two valence bands. The lower conduction band and the upper valence band touch

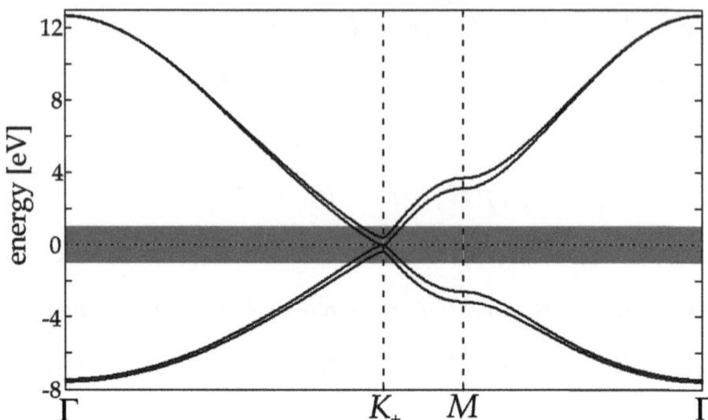

Fig. 2.2 The band structure of bilayer graphene resulting from Eq. (2.4) with the Hamiltonian and overlap matrices $\hat{H}$ and $\hat{S}$ as in Eq. (2.6) presented for high-symmetry directions in the Brillouin zone as shown in Fig. 2.1b. The values of the parameters used are: $\gamma_0 = 3.1\,\text{eV}$, $\gamma_1 = 0.4\,\text{eV}$, $\gamma_3 = 0.15\,\text{eV}$, $\gamma_4 = 0.1\,\text{eV}$, $\gamma_n = 0.05\,\text{eV}$, $s_0 = 0.1$, $s_1 = -0.05$, $\epsilon_i = 0$ for all i, and $a = 2.46\,\text{Å}$. The range of energies important in experimental setups modelled theoretically in this work is shown on *red* background

exactly at the K point. The position of this touching point at the energy scale denotes half-filling of the bands and is called the neutrality point. It is usually treated as the zero of the energy scale. In the neutral structure with the valence bands completely filled, the Fermi surface consists only of points. Any shift of the chemical potential results in the creation of separate Fermi lines around each of the K points. Due to this behaviour of the Fermi surface, the K points are often refered to as *valleys*. We point out that for most experiments and potential applications, only the part of the dispersion relatively close to the neutrality point (~ 1 eV) is important and interesting. Indeed, a proper description of this part of the band structure forms the basis for the understanding and theoretical modelling of spectroscopic experiments presented in this thesis. We will, therefore, investigate it in more detail in the following sections.

2.2.2 Approximation for Hopping Elements

We now want to look closer at the electronic dispersion for energies relevant to most experiments, that is, the energies of up to ~ 1 eV from the neutrality point. This range of energies is marked with red background in Fig. 2.2. For such energy, the relevant regions in momentum space are the vicinities of the six corners of the Brillouin zone. To describe electronic dispersion around a local minima at the K points, we shift the coordinate system in reciprocal (momentum) space from the Γ point to the K_ξ point. We write the electron wave vector as $\boldsymbol{k} = \boldsymbol{K}_\xi + \frac{\boldsymbol{p}}{\hbar}$, where the electronic momentum $\boldsymbol{p}$ is now measured from the centre of the valley K_ξ. The geometrical factor $f(\boldsymbol{k})$ from Eq. (2.5), expanded up to the second order in $\boldsymbol{p}$, reads

$$f(\boldsymbol{k}) \approx -\frac{\sqrt{3}a}{2\hbar}\left(\xi p_x - i p_y\right) + \frac{a^2}{8\hbar^2}\left(p_x + i p_y\right)^2 . \tag{2.8}$$

We introduce some new parameters, namely velocities $v = \frac{\sqrt{3}a\gamma_0}{2\hbar}$, $v_3 = \frac{\sqrt{3}a\gamma_3}{2\hbar}$ and $v_4 = -\frac{\sqrt{3}a\gamma_4}{2\hbar}$, constant $\eta = \frac{a^2\gamma_0}{8\hbar^2} \equiv \frac{v^2}{6\gamma_0}$, as well as operators, $\hat{\pi} = p_x + ip_y$ and $\hat{\pi}^\dagger = p_x - ip_y$. We also neglect at this stage the overlap integrals s_0 and s_1, and hence obtain only a unit matrix on the right hand side of Eq. (2.4). Our basis now consists of 8 functions $\phi_{p,\xi,i}(r)$ (4 for each valley). However, for the cases considered in this thesis, the valleys can be regarded as independent (we do not consider any valley-connecting processes) and it is usually enough to explicitly write down only the electronic Hamiltonian $\hat{H}_\xi$ for one valley K_ξ,

$$
\hat{H}_\xi = \xi \begin{pmatrix} 0 & v_3\hat{\pi} & v_4\hat{\pi}^\dagger & v\hat{\pi}^\dagger \\ v_3\hat{\pi}^\dagger & 0 & v\hat{\pi} & v_4\hat{\pi} \\ v_4\hat{\pi} & v\hat{\pi}^\dagger & 0 & \xi\gamma_1 \\ v\hat{\pi} & v_4\hat{\pi}^\dagger & \xi\gamma_1 & 0 \end{pmatrix} - \eta \begin{pmatrix} \frac{6\gamma_n}{\gamma_0}p^2 & \frac{v_3}{v}(\hat{\pi}^\dagger)^2 & \frac{v_4}{v}\hat{\pi}^2 & \hat{\pi}^2 \\ \frac{v_3}{v}\hat{\pi}^2 & \frac{6\gamma_n}{\gamma_0}p^2 & (\hat{\pi}^\dagger)^2 & \frac{v_4}{v}(\hat{\pi}^\dagger)^2 \\ \frac{v_4}{v}(\hat{\pi}^\dagger)^2 & \hat{\pi}^2 & \frac{6\gamma_n}{\gamma_0}p^2 & 0 \\ (\hat{\pi}^\dagger)^2 & \frac{v_4}{v}\hat{\pi}^2 & 0 & \frac{6\gamma_n}{\gamma_0}p^2 \end{pmatrix}.
$$

$$(2.9)$$

In the Hamiltonian above, we have for now neglected the on-site energies ϵ_i, which are discussed in detail in Sect. 2.2.3. The order of the basis functions is[2] $(\phi_{+,A1}, \phi_{+,B2}, \phi_{+,A2}, \phi_{+,B1})^T$ in the K_+ and $(\phi_{-,B2}, \phi_{-,A1}, \phi_{-,B1}, \phi_{-,A2})^T$ in the K_- valley.

The same Hamiltonian can be obtained using the 'k · p' approximation (see for example [11, 12] for detailed derivation). In this scheme, as a basis set we use functions $\tilde{\phi}_{p,\xi,i}(r)$ constructed from $\phi_{k,i}(r)$, Eq. (2.1), calculated exactly in the centre of the valley K_ξ and a plane wave envelope function which varies slowly at the distance of the order of the lattice constant a [13, 14]:

$$
\tilde{\phi}_{p,\xi,i}(r) \equiv e^{\frac{i}{\hbar}p\cdot r}\phi_{K_\xi,i}(r) \equiv \frac{1}{\sqrt{N}}e^{\frac{i}{\hbar}p\cdot r}\sum_{R_0} e^{iK_\xi\cdot(R_0+R_i)}\varphi(r - R_0 - R_i).
$$

Comparing functions $\tilde{\phi}_{p,\xi,i}(r)$ and $\phi_{K_\xi+\frac{p}{\hbar},i}(r)$, we can intuitively see why the Hamiltonians in both approximations take the same form. Both functions take similar values for $r \approx R_0 - R_i$, whereas in other regions the π orbital $\varphi(r - R_0 - R_i)$ ensures that they both quickly decay, rendering the phase factors unimportant.

The electronic band structure resulting from the Hamiltonian (2.9) is shown for both valleys in Fig. 2.3 for the energy range $3.5\gamma_1$ away from the neutrality point. On this scale, all bands look approximately parabolic very close to the center of the valley and linear further away. The former is not exactly true for the bands shown in yellow (later referred to as the low-energy bands), as shown in Sect. 2.3. The bands marked in red (in what follows called the high-energy or split bands) are shifted away from the neutrality point by approximately the interlayer coupling, $\gamma_1 \sim 0.4\,\text{eV}$ [15–23] in each direction. The velocity $v \sim 10^6\,\text{m/s}$ [18–20, 23] determines the slope of the

---

[2] For brevity, we omit the momentum index p and explicit dependence of the basis functions $\phi_{p,\xi,i}$ on r.

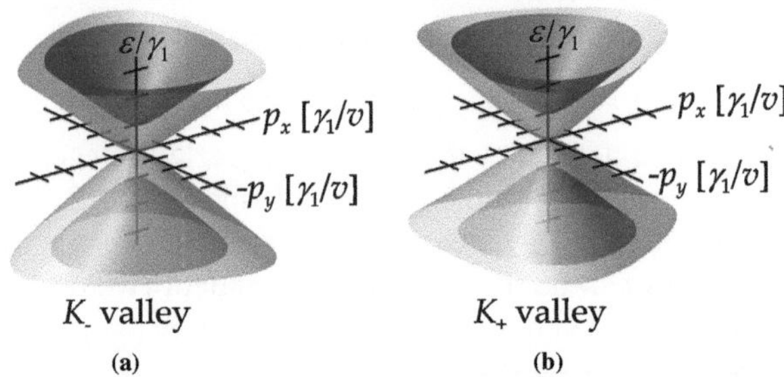

Fig. 2.3 The band structure of bilayer graphene in the vicinity of both (**a**) K_- and (**b**) K_+ valleys, obtained within the linear approximation, Eq. (2.9). The values of the parameters used are: $\gamma_0 = 3.1\,\text{eV}$, $\gamma_1 = 0.4\,\text{eV}$, $\gamma_3 = 0.2\,\text{eV}$, $\gamma_4 = 0.1\,\text{eV}$, $\gamma_n = 0.1\,\text{eV}$, $a = 2.46\,\text{Å}$. No on-site asymmetries are included (that is, $\epsilon_i = 0$ for all i). For both valleys, the low-energy and high-energy bands are shown in *yellow* and *red*, respectively

linear parts of the bands. The isoenergetic lines create circles, which in the case of the low-energy bands are trigonally warped. This warping is the effect of the velocity $v_3 \sim 0.1v$ [20, 23] as well as terms quadratic in the momentum $\boldsymbol{p}$. The remaining velocity, v_4, breaks the electron-hole symmetry. It renormalizes somewhat the slopes of the linear parts and its effect is opposite in the conduction and valence bands. The next-nearest neighbour coupling γ_n also breaks the electron-hole symmetry. The parameters v, γ_1 and v_3 are the most important in the description of the electronic dispersion around the valleys. At low energies, the deviation from the electron-hole symmetric situation is negligible for most cases. Also, it is difficult to experimentally separate the contributions of γ_4, γ_n or even s_0 to the electron-hole asymmetry.

2.2.3 Symmetry-Breaking Asymmetries in the on-Site Energies

Up to this point, we considered the kinetic energies of electrons on different atomic sites (the terms ϵ_i in Eq. 2.7a) to be equal. In other words, all the carbon atoms in the lattice were chemically equivalent. However, that is obviously not the case as the environment of the dimer atoms, $B1$ and $A2$, is definitely different than the surroundings of the atoms $A1$ and $B2$. In general, three parameters are needed to account for differences between our four atomic sites. The (not unique) definitions we use here are:

$$\Delta_{AB} = \frac{1}{2}[(\epsilon_{A1} + \epsilon_{A2}) - (\epsilon_{B1} + \epsilon_{B2})]; \tag{2.10a}$$

$$\Delta = \frac{1}{2}[(\epsilon_{A1} + \epsilon_{B2}) - (\epsilon_{B1} + \epsilon_{A2})]; \tag{2.10b}$$

$$u = \frac{1}{2}[(\epsilon_{A1} + \epsilon_{B1}) - (\epsilon_{A2} + \epsilon_{B2})]; \tag{2.10c}$$

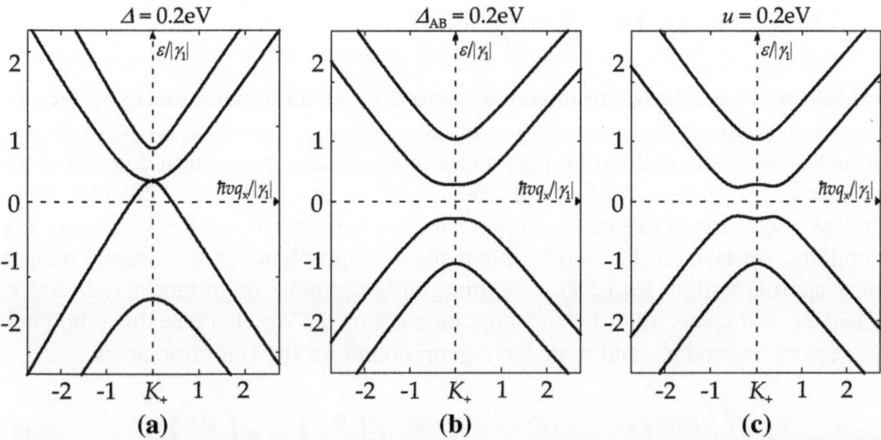

Fig. 2.4 Cuts through the electronic dispersion around the K_+ valley along the p_x axis in the presence of (**a**) $\Delta = 0.2\,\text{eV}$, (**b**) $\Delta_{AB} = 0.2\,\text{eV}$, (**c**) $u = 0.2\,\text{eV}$. For illustrative purposes, we use large values of asymmetries. In reality, only the interlayer asymmetry u can be of the order of $100\,\text{meV}$ [25, 26]. The values of other parameters used are: $\gamma_0 = 3\,\text{eV}$, $\gamma_1 = 0.35\,\text{eV}$, $\gamma_3 = 0.15\,\text{eV}$, $\gamma_4, \gamma_n = 0\,\text{eV}$, $a = 2.46\,\text{Å}$. Figure reprinted from Ref. [27], Copyright (2010), with permission from IOP Publishing

$$\epsilon_{A1} = \frac{1}{2}(u + \Delta + \Delta_{AB}); \qquad \epsilon_{B1} = \frac{1}{2}(u - \Delta - \Delta_{AB}); \qquad (2.10\text{d})$$

$$\epsilon_{A2} = \frac{1}{2}(-u - \Delta + \Delta_{AB}); \qquad \epsilon_{B2} = \frac{1}{2}(-u + \Delta - \Delta_{AB}); \qquad (2.10\text{e})$$

Then, Δ_{AB} describes the difference between on-site energies of A and B sublattice sites on each layer. We call it intralayer asymmetry. It may be influenced, especially in the bottom layer, by the underlying substrate. The next parameter, Δ, accounts for an energy difference between dimer and non-dimer sites. Finally, u characterizes the interlayer asymmetry between the two layers. This asymmetry can be significantly changed by doping the sample [15] or even continously varied with external gates [24]. This effect is discussed in more depth in Sect. 4.3.1. To show the influence of each of the asymmetries on the band structure, we add them separately to the tight-binding model and plot in Fig. 2.4 cuts through the electronic dispersion around the K_+ point along the p_x axis. The intralayer asymmetry does not open any gaps in the electronic spectrum but breaks the electron-hole symmetry. The dimer/non-dimer asymmetry Δ_{AB} opens a gap and preserves the electron-hole symmetry. The interlayer asymmetry u preserves the electron-hole symmetry and also leads to the opening of a gap in the spectrum. However, we point out the characteristic 'Mexican-hat-like' features in the shape of the low-energy bands in the vicinity of the gap which appear due to u (Fig. 2.4c).

2.3 The Effective Two-Band Model

In this section, we derive an effective low-energy Hamiltonian describing the two degenerate bands in the close vicinity of the valley K_ξ. This analysis was first performed by McCann and Fal'ko [28] and proved extremely useful in describing the low-energy properties of bilayer graphene (see for example Refs. [28–33]). For a detailed discussion of this effective two-band approximation, see Refs. [27, 34]. For simplicity, we take as the starting point the eigenproblem for the Hamiltonian in linear approximation, Eq. (2.9), containing only the most important parameters v, γ_1 and v_3, and terms up to linear in the momentum $\boldsymbol{p}$. We also introduce the Pauli matrices σ_x, σ_y and σ_z, and write the eigenproblem for the Hamiltonian as

$$\begin{pmatrix} \xi v_3(\sigma_x p_x - \sigma_y p_y) & \xi v(\boldsymbol{\sigma} \cdot \boldsymbol{p}) \\ \xi v(\boldsymbol{\sigma} \cdot \boldsymbol{p}) & \sigma_x \gamma_1 \end{pmatrix} \begin{pmatrix} \psi_1 \\ \psi_2 \end{pmatrix} = \epsilon \begin{pmatrix} \psi_1 \\ \psi_2 \end{pmatrix}, \qquad (2.11)$$

where ψ_1 and ψ_2 denote two-component vectors which form together the four-component electronic eigenstate ψ_j from (2.4). Let us use the second row of (2.11) to express ψ_2 in terms of ψ_1 and ϵ and substitute it into the first row to obtain,

$$\xi v_3 \left(\sigma_x p_x - \sigma_y p_y\right) \psi_1 + v^2 \left(\boldsymbol{\sigma} \cdot \boldsymbol{p}\right) \left[\epsilon - \sigma_x \gamma_1\right]^{-1} \left(\boldsymbol{\sigma} \cdot \boldsymbol{p}\right) \psi_1 = \epsilon \psi_1.$$

For low energies, $\epsilon \ll \gamma_1$, we get $\left[\epsilon - \sigma_x \gamma_1\right]^{-1} \approx -\frac{\sigma_x}{\gamma_1}$ and

$$\left\{ -\frac{v^2}{\gamma_1} \left[\sigma_x \left(p_x^2 - p_y^2 \right) + 2\sigma_y p_x p_y \right] + \xi v_3 \left(\sigma_x p_x - \sigma_y p_y \right) \right\} \psi_1$$
$$\equiv \hat{\boldsymbol{H}}_{\text{eff}} \psi_1 = \epsilon \psi_1. \qquad (2.12)$$

The above effective Hamiltonian describes the electronic dispersion for energies close to the neutrality point while neglecting the split bands. The basis of $\hat{\boldsymbol{H}}_{\text{eff}}$ is $(\phi_{+,A1}, \phi_{+,B2})^T$ at the K_+ and $(\phi_{-,B2}, \phi_{-,A1})^T$ at the K_- valley. The resulting electronic dispersion,

$$\epsilon_\xi = \pm \left[\left(\frac{v^2}{\gamma_1} \right)^2 p^4 + v_3^2 p^2 - \frac{2\xi v^2 v_3}{\gamma_1} p^3 \cos 3\varphi \right]^{\frac{1}{2}}, \qquad (2.13)$$

where $p^2 = p_x^2 + p_y^2$ and $\arctan \varphi = \frac{p_y}{p_x}$, is shown in Fig. 2.5a. The trigonally warped isoenergetic line undergoes a splitting into four pockets at the energy $\epsilon_{\text{LT}} = \pm \frac{\gamma_1}{4} \left(\frac{v_3}{v} \right)^2$, $|\epsilon_{\text{LT}}| \sim 1\,\text{meV}$ (a so called Lifshitz transition [35]), see Fig. 2.5b. However, characteristic values for this transition energy ϵ_{LT} and momentum $p_{\text{LT}} \sim \frac{\gamma_1 v_3}{v^2}$ are below the resolution of any of spectroscopies considered in the following chapters. Hence, the only importance of v_3 for our considerations is its role as the main source of trigonal warping for the isoenergetic lines at low energies.

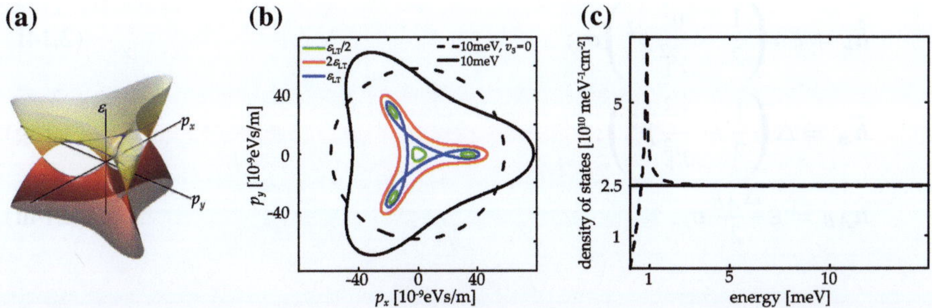

Fig. 2.5 **a** The electronic dispersion of bilayer graphene at very low energies $\epsilon \sim 1$ meV around the valley K_+. The blue contour represents the isoenergetic line at the Lifshitz transition energy $\epsilon_{\mathrm{LT}} = \pm\frac{\gamma_1}{4}\left(\frac{v_3}{v}\right)^2$. **b** The isoenergetic lines around the valley K_+ for energies $\epsilon_{\mathrm{LT}}/2$ (*green solid line*), ϵ_{LT} (*blue solid line*) and $2\epsilon_{\mathrm{LT}}$ (*red solid line*). Also shown are the isoenergetic line at the energy 10 meV from the neutrality point (black solid line) and its shape if the trigonal warping due to v_3 is neglected (*black dashed line*). **c** The density of states given by Eq. (2.13) for $v_3 = 0.1v$ (*dashed line*) and $v_3 = 0$ (*solid line*). The latter corresponds to a purely parabolic bottom of the band

For these energies, trigonal warping does not significantly affect the density of states (DOS), as shown in Fig. 2.5c, where the comparison between the density of states of a purely parabolic band in the case of $v_3 = 0$ (black solid line) and of the electronic dispersion given by Eq. (2.13) (black dashed line), is shown. The peak in the DOS corresponds to the Lifshitz transition. However, for energies $\epsilon > 5$ meV the density of states already closely follows constant density of states for a parabolic band, although the isoenergetic lines may still be significantly noncircular (Fig. 2.5b).

The procedure applied above to obtain the low-energy description of electrons can be easily generalised to include all other terms appearing in the four-band Hamiltonian in linear approximation, Eq. (2.9), as well as the on-site asymmetries from Sect. 2.2.3. We then obtain [27]

$$\hat{\boldsymbol{H}}_{\mathrm{eff}} = \hat{\boldsymbol{H}}_0 + \hat{\boldsymbol{h}}_w + \hat{\boldsymbol{h}}_4 + \hat{\boldsymbol{h}}_n + \hat{\boldsymbol{h}}_u + \hat{\boldsymbol{h}}_\Delta + \hat{\boldsymbol{h}}_{AB}, \tag{2.14a}$$

where

$$\hat{\boldsymbol{H}}_0 = -\frac{v^2}{\gamma_1}\left[\sigma_x\left(p_x^2 - p_y^2\right) + 2\sigma_y p_x p_y\right]; \tag{2.14b}$$

$$\hat{\boldsymbol{h}}_w = \xi v_3\left(\sigma_x p_x - \sigma_y p_y\right) - \eta\frac{v_3}{v}\left[\sigma_x\left(p_x^2 - p_y^2\right) + 2\sigma_y p_x p_y\right]; \tag{2.14c}$$

$$\hat{\boldsymbol{h}}_4 = 2\frac{\gamma_4 v^2}{\gamma_0\gamma_1}p^2; \tag{2.14d}$$

$$\hat{\boldsymbol{h}}_n = -\frac{\gamma_n v^2}{\gamma_0^2}p^2; \tag{2.14e}$$

$$\hat{\boldsymbol{h}}_u = \xi u \left(\frac{1}{2} - \frac{v^2}{\gamma_1^2} p^2 \right) \sigma_z; \tag{2.14f}$$

$$\hat{\boldsymbol{h}}_\Delta = \Delta \left(\frac{1}{2} - \frac{v^2}{\gamma_1^2} p^2 \right); \tag{2.14g}$$

$$\hat{\boldsymbol{h}}_{AB} = \xi \frac{\Delta_{AB}}{2} \sigma_z. \tag{2.14h}$$

References

1. P.R. Wallace, The band theory of graphite. Phys. Rev. **71**, 622 (1947)
2. J.W. McClure, Band structure of graphite and de Haas-van Alphen effect. Phys. Rev. **108**, 612 (1957)
3. J.C. Slonczewski, P.R. Weiss, Band structure of graphite. Phys. Rev. **109**, 272 (1958)
4. J.W. McClure, Theory of diamagnetism of graphite. Phys. Rev. **119**, 606 (1960)
5. R. Saito, G. Dresselhaus, M.S. Dresselhaus, *Physical Properties of Carbon Nanotubes* (Imperial College Press, London, 1998)
6. C. Bena, G. Montambaux, Remarks on the tight-binding model of graphene. New J. Phys. **11**, 095003 (2009)
7. J.D. Bernal, The structure of graphite. Proc. R. Soc. A **106**, 749 (1924)
8. M.S. Dresselhaus, G. Dresselhaus, Intercalation compounds of graphite. Adv. Phys. **30**, 139 (1981)
9. F. Varchon, R. Feng, J. Hass, X. Li, B. Ngoc Nguyen, C. Naud, P. Mallet, J.-Y. Veuillen, C. Berger, E.H. Conrad, L. Magaud, Electronic structure of epitaxial graphene layers on SiC: effect of the substrate. Phys. Rev. Lett. **99**, 126805 (2007)
10. S.B. Trickey, F. Müller-Plathe, G.H.F. Diercksen, Interplanar binding and lattice relaxation in a graphite dilayer. Phys. Rev. B **45**, 4460 (1992)
11. H. Ajiki, T. Ando, Electronic states of carbon nanotubes. J. Phys. Soc. Jpn. **62**, 1255 (1993)
12. T. Ando, Theory of electronic states and transport in carbon nanotubes. J. Phys. Soc. Jpn. **74**, 777 (2005)
13. J.M. Luttinger, W. Kohn, Motion of electrons and holes in perturbed periodic fields. Phys. Rev. **97**, 869 (1955)
14. D.P. DiVincenzo, E.J. Mele, Self-consistent effective-mass theory for intralayer screening in graphite intercalation compounds. Phys. Rev. B **29**, 1685 (1984)
15. T. Ohta, A. Bostwick, T. Seyller, K. Horn, E. Rotenberg, Controlling the electronic structure of bilayer graphene. Science **313**, 951 (2006)
16. L.M. Zhang, Z.Q. Li, D.N. Basov, M.M. Fogler, Z. Hao, M.C. Martin, Determination of the electronic structure of bilayer graphene from infrared spectroscopy. Phys. Rev. B **78**, 235408 (2008)
17. A.B. Kuzmenko, E. van Heumen, D. van der Marel, P. Lerch, P. Blake, K.S. Novoselov, A.K. Geim, Infrared spectroscopy of electronic bands in bilayer graphene. Phys. Rev. B **79**, 115441 (2009)
18. A.B. Kuzmenko, I. Crassee, D. van der Marel, P. Blake, K.S. Novoselov, Determination of the gate-tunable band gap and tight-binding parameters in bilayer graphene using infrared spectroscopy. Phys. Rev. B **80**, 165406 (2009)
19. Z.Q. Li, E.A. Henriksen, Z. Jiang, Z. Hao, M.C. Martin, P. Kim, H.L. Stormer, D.N. Basov, Band structure asymmetry of bilayer graphene revealed by infrared spectroscopy. Phys. Rev. Lett. **102**, 037403 (2009)

20. L.M. Malard, J. Nilsson, D.C. Elias, J.C. Brant, F. Plentz, E.S. Alves, A.H. Castro Neto, M.A. Pimenta, Probing the electronic structure of bilayer graphene by Raman scattering. Phys. Rev. B **76**, 201401(R) (2007)
21. J. Yan, E.A. Henriksen, P. Kim, A. Pinczuk, Observation of anomalous phonon softening in bilayer graphene. Phys. Rev. Lett. **101**, 136804 (2008)
22. A. Das, B. Chakraborty, S. Piscanec, S. Pisana, A.K. Sood, A.C. Ferrari, Phonon renormalization in doped bilayer graphene. Phys. Rev. B **79**, 155417 (2009)
23. D.L. Mafra, L.M. Malard, S.K. Doorn, H. Htoon, J. Nilsson, A.H. Castro Neto, M.A. Pimenta, Observation of the Kohn anomaly near the K point of bilayer graphene. Phys. Rev. B **80**, 241414(R) (2009)
24. J.B. Oostinga, H.B. Heersche, X.L. Liu, A.F. Morpurgo, L.M.K. Vandersypen, Gate-induced insulating state in bilayer graphene devices. Nat. Mater. **7**, 151 (2008)
25. Y. Zhang, T.-T. Tang, C. Girit, Z. Hao, M.C. Martin, A. Zettl, M.F. Crommie, Y.R. Shen, F. Wang, Direct observation of a widely tunable bandgap in bilayer graphene. Nature **459**, 820 (2009)
26. K.F. Mak, C.H. Lui, J. Shan, T.F. Heinz, Observation of an electric-field-induced band gap in bilayer graphene by infrared spectroscopy. Phys. Rev. Lett. **102**, 256405 (2009)
27. M. Mucha-Kruczyński, E. McCann, V.I. Fal'ko, Electron-hole asymmetry and energy gaps in bilayer graphene. Semicond. Sci. Technol. **25**, 033001 (2010)
28. E. McCann, V.I. Fal'ko, Landau level degeneracy and quantum hall effect in a graphite bilayer. Phys. Rev. Lett. **96**, 086805 (2006)
29. D.S.L. Abergel, V.I. Fal'ko, Optical and magneto-optical far-infrared properties of bilayer graphene. Phys. Rev. B **75**, 155430 (2007)
30. M.I. Katsnelson, K.S. Novoselov, A.K. Geim, Chiral tunnelling and the Klein paradox in graphene. Nat. Phys. **2**, 620 (2006)
31. K. Kechedzhi, E. McCann, V. Fal'ko, B. Altshuler, Influence of trigonal warping on interference effects in bilayer graphene. Phys. Rev. Lett. **98**, 176806 (2007)
32. K.S. Novoselov, E. McCann, S.V. Morozov, V.I. Fal'ko, M.I. Katsnelson, U. Zeitler, D. Jiang, F. Schedin, A.K. Geim, Unconventional quantum Hall effect and Berry's phase of 2pi in bilayer graphene. Nat. Phys. **2**, 177 (2006)
33. M.I. Katsnelson, Minimal conductivity in bilayer graphene. Eur. Phys. J. B **52**, 151 (2006)
34. E. McCann, D.S.L. Abergel, V.I. Fal'ko, Electrons in bilayer graphene. Solid State Commun. **143**, 110 (2007)
35. I. Lifshitz, Anomalies of electron characteristics of a metal in the high pressure region. Sov. Phys. J. Exp. Theor. Phys. **11**, 1130 (1960)

Chapter 3
Angle-Resolved Photoemission Spectroscopy

The angle-resolved photoemission spectroscopy (ARPES) is a powerful experimental tool based on the photoelectric effect, first observed by Hertz more than 120 years ago [1] and explained by Einstein at the beginning of the previous century with the help of the then novel idea of photons, quanta of electromagnetic radiation [2]. In the photoelectric effect, absorption of a sufficiently energetic incident photon with energy ω ejects an electron with the initial energy ϵ_p corresponding to the momentum $\boldsymbol{p} = \hbar\boldsymbol{k}$ from the sample into the vacuum. In ARPES, such ejected electrons (called photoelectrons) with kinetic energy ϵ_e are detected with the help of a hemispherical detectors, so that both their energy and momentum $\boldsymbol{p}_e$ can be identified. The modulus of the latter is given by $p_e = \sqrt{2m_e\epsilon_e}$ (m_e is the electron mass), while its components,

$$
(p_e)_x = \sqrt{2m_e\epsilon_e}\,\cos\phi\,\sin\theta;
$$
$$
(p_e)_y = \sqrt{2m_e\epsilon_e}\,\sin\phi\,\sin\theta;
$$
$$
p_e^{\perp} = \sqrt{2m_e\epsilon_e}\,\cos\theta; \tag{3.1}
$$

where ϕ and θ are the azimuthal and polar angles of detection, respectively. Knowledge of electronic states in the sample is gained with the help of two conservation laws: (I) conservation of the energy in the whole process puts a constraint on ϵ_p, while (II) conservation of the in-plane-momentum resulting from in-plane crystallic periodicity yields some information about the electron momentum in the crystal, $\boldsymbol{p}$. For bulk materials, the angular distributions of measured photoelectrons as a function of ϵ_p are difficult to analyse, due to the lack of sufficient restraint on the out-of-plane component of $\boldsymbol{p}$. For layered or (quasi-)two-dimensional systems, however, as long as the incident radiation is monochromatic, those distributions represent direct connection to the constant-energy contours of the band structure of the material.

Because of its layered nature, graphite has been an object of extensive ARPES studies in the past [3–9]. Poor angular and energy resolutions resulted in little data concerning the π bands in the vicinity of the K points, although significant variation in the intensity from states on the same isoenergetic line has been noticed [8, 9]. Much more sophisticated equipment was available at the moment when monolayer

M. Mucha-Kruczyński, *Theory of Bilayer Graphene Spectroscopy*,
Springer Theses, DOI: 10.1007/978-3-642-30936-6_3,
© Springer-Verlag Berlin Heidelberg 2013

graphene was isolated. Hence, ARPES was the method of choice for numerous investigations of the electronic band structure of two-dimensional graphene systems. It has been used to examine the epitaxial growth and confirm the graphene-like linear dispersion relation for electrons in carbon layers on SiC [10, 11], Ni [12–15], Ir [16] or Ru [17]. At the same time, high resolution allowed detailed examination of the deviations from this linearity [18–21], which stimulated numerous theoretical considerations of the band renormalisation due to many-body interactions [22–27] and some controversy on the possibility of the substrate-induced sublattice symmetry breaking in the monolayer grown on SiC [21, 28–30]. ARPES has been used to show for the first time, the appearance of the electric-field-induced gap in the spectrum of bilayer graphene [31]. In the same work, the magnitude of the interlayer coupling γ_1 has been extracted. It has been further employed to investigate the effects of molecular doping of monolayer and bilayer graphene [32, 33]. The ARPES studies have been then extended to tri- and fourlayer graphene systems [34]. Review of the photoemission studies of graphene systems grown on SiC can be found in Refs. [35, 36].

In this chapter, we aim to describe the angular distribution of the angle-resolved photoemission spectroscopy intensity patterns for bilayer graphene. We first introduce a simple theoretical model of the photoemission process, based on the idea of multiple-source interference of electronic Bloch waves. More detailed reviews of the theoretical background of ARPES can be found, for example, in Refs. [37, 38]. We then use our model to obtain low-energy angular distributions of ARPES intensity for monolayer graphene and demonstrate that they are a manifestation of what has been recently branded as electronic *chirality* and what is common to all graphene-layered systems, including graphite [9, 39]. Afterwards, we show that for bilayer graphene specifically, the anisotropy of the constant-energy maps may be used to extract information about the magnitude and sign of interlayer coupling parameters and about symmetry breaking inflicted on a bilayer by the underlying substrate. Our investigation was first published in Ref. [40].

3.1 ARPES as Quantum Young's Experiment

We consider here the following photoemission process: an incoming photon with energy $\omega > W$, where W is the work function of the material, is absorbed by an electron in the momentum state k with energy ϵ_k. This electron receives thus, enough energy to overcome the energetic barrier, described by the work function W, and leave the material. It is then detected as having energy ϵ_e and momentum p_e, connected by Eq. 3.1. The energy conservation in the whole process can be expressed as

$$\omega + \epsilon_k = W + \epsilon_e. \tag{3.2}$$

We treat the electron leaving the material as a simple case of a wave passing a potential step. It follows then from the periodicity of the sample in the plane, that the in-plane component of the momentum is conserved,

$$\hbar(\boldsymbol{k} + \boldsymbol{G}) = p_e^{\parallel},\tag{3.3}$$

where $\boldsymbol{p} = \hbar\boldsymbol{k}$, $\boldsymbol{G} = m_1\boldsymbol{b}_1 + m_2\boldsymbol{b}_2$ is the reciprocal lattice vector and we choose the z axis to correspond to the direction perpendicular to the plane of our sample and $(p_e^{\parallel})^2 = \sqrt{p_e^2 - (p_e^{\perp})^2}$; $p_e^{\perp} = (0, 0, p_e^{\perp})$.

The ARPES intensity is proportional to the modulus square of the transition amplitude between the initial and final states of system under the perturbation caused by incoming radiation. We treat the latter with the perturbative Hamiltonian

$$\hat{H}_{\text{int}} = -\frac{e}{2m_e}\left(\boldsymbol{A}\cdot\hat{\boldsymbol{p}} + \hat{\boldsymbol{p}}\cdot\boldsymbol{A}\right) = -\frac{e}{m_e}\boldsymbol{A}\cdot\hat{\boldsymbol{p}},\tag{3.4}$$

where $\boldsymbol{A}$ is the electromagnetic vector potential of the incoming radiation and $\hat{\boldsymbol{p}}$ is the electron momentum operator. We neglect many-body interactions and as the initial state take the single-electron Bloch wave state in the general form, Eq. (2.2),

$$\Psi(\boldsymbol{r}) = \sum_i C_i \phi_{k,i}(\boldsymbol{r}),\tag{3.5}$$

where we have for now dropped the index j and consider a single band. As for the final state of the electron, $\Psi_e(\boldsymbol{r})$, we approximate it with a plane wave,

$$\Psi_e(\boldsymbol{r}) \propto \exp\left(\frac{i}{\hbar}\boldsymbol{p}_e\cdot\boldsymbol{r}\right).\tag{3.6}$$

We are interested in the angular distribution of ARPES probing the low-energy electronic states in the vicinity of the valleys, not the absolute value of the intensity. Hence, neglecting prefactors not contributing to the angle-dependence, we express the ARPES intensity from electron states in a given band as

$$I \propto \left|\langle e^{\frac{i}{\hbar}\boldsymbol{p}_e\cdot\boldsymbol{r}}|\boldsymbol{A}\cdot\boldsymbol{p}|\sum_{i,\boldsymbol{R}_0} C_i e^{i\boldsymbol{k}\cdot(\boldsymbol{R}_0+\boldsymbol{R}_i)}\varphi(\boldsymbol{r} - \boldsymbol{R}_0 - \boldsymbol{R}_i)\rangle\right|^2$$
$$\times\ \delta\left(\epsilon_e + W - \epsilon_p - \omega\right).\tag{3.7}$$

At the same time, we expect the patterns to reflect the shape of the isoenergetic lines around the valleys. The radius of the area in the reciprocal space around a single valley important for our considerations is less than ten per cent of the $\Gamma - K$ distance. Also, the shapes of the patterns for a specific energy should resemble trigonally warped circles with all important angle-dependent features contained within a narrow ring in the reciprocal (or momentum) space. We assume, that for this narrow range of momenta, the result of the perturbation operator $\boldsymbol{A}\cdot\hat{\boldsymbol{p}}$ acting on the initial state is a smooth, slowly varying function of momentum and approximate it with an irrelevant constant. This assumption is justified, as the energy of incoming photons used in experiments is of the order of 50–150 eV [11, 17, 21, 28, 31, 34, 35] translating into the lower bound on the photon's wavelength $\lambda \gtrsim 5$ nm, corresponding to about twenty lattice constants. This means that the incoming electromagnetic wave does not distinguish details of the electron Bloch state (for example the structure of the atomic $2p_z$ orbital). Therefore, the intensity I can be related to

$$I \propto \left| \sum_i C_i e^{\frac{i}{\hbar}(\boldsymbol{p}-\boldsymbol{p}_e)\cdot\boldsymbol{R}_i} \sum_{\boldsymbol{R}_0} e^{\frac{i}{\hbar}(\boldsymbol{p}-\boldsymbol{p}_e)\cdot\boldsymbol{R}_0} \langle e^{\frac{i}{\hbar}\boldsymbol{p}_e\cdot\boldsymbol{r}'} | \varphi(\boldsymbol{r}') \rangle \right|^2 \delta\left(\epsilon_e + W - \epsilon_{\boldsymbol{p}} - \omega\right),$$

where we introduced new position vector $\boldsymbol{r}' = \boldsymbol{r} - \boldsymbol{R}_0 - \boldsymbol{R}_i$. The sum over the two-dimensional lattice vectors $\boldsymbol{R}_0$ of the phase factors leads to the Dirac delta expressing conservation of the in-plane momentum, Eq. 3.3. The integral over $\boldsymbol{r}'$ is the Fourier image of the atomic $2p_z$ orbital $\varphi(\boldsymbol{r}')$, which we denote by $\varphi(\boldsymbol{p}_e)$. We obtain

$$I \propto |\varphi(\boldsymbol{p}_e)|^2 \left| \sum_i C_i e^{-i\boldsymbol{G}\cdot\boldsymbol{R}_i} e^{-\frac{i}{\hbar}\boldsymbol{p}_e^{\perp}\cdot\boldsymbol{R}_i} \right|^2 \delta\left(\epsilon_e + W - \epsilon_{\boldsymbol{p}} - \omega\right). \tag{3.8}$$

Let us now consider the Fourier transform $\varphi(\boldsymbol{p}_e)$ as a function written in spherical coordinates, $\varphi(\boldsymbol{p}_e) \equiv \varphi(p_e, \phi, \theta)$. Just like the $2p_z$ orbital in the real space, its Fourier transform $\varphi(\boldsymbol{p}_e)$ has rotational symmetry in the $p_x - p_y$ plane and does not depend on the azimuthal angle ϕ. For a given energy of the incoming photons ω and material specific work function W, to resolve the constant-energy maps of the ARPES intensity for energy $\epsilon_{\boldsymbol{p}}$, one only needs to look at photoelectrons with well specified energy $\epsilon_e = \omega - W + \epsilon_{\boldsymbol{p}}$ and thus the modulus of the momentum p_e. Finally, as we concentrate on a small area in the momentum space around the valleys, the resulting change in the polar angle θ is going to be small. For the momentum states under consideration, $\varphi(\boldsymbol{p}_e)$ is essentially a constant. Therefore, we neglect the prefactor $\left|\varphi(\boldsymbol{p}_e)\right|^2$. We will also first treat the lattice as strictly two-dimensional, which means that $e^{-\frac{i}{\hbar}\boldsymbol{p}_e^{\perp}\cdot\boldsymbol{R}_i} = 1$. We end with the expression

$$I \propto \left| \sum_i C_i e^{-i\boldsymbol{G}\cdot\boldsymbol{R}_i} \right|^2 \delta(\epsilon_e + W - \epsilon_{\boldsymbol{k}} - \omega). \tag{3.9}$$

In the formula above, the ARPES intensity pattern arises as a result of the interference of the photoelectron waves originating from all atomic sites within the unit cell. The contribution of the i-th site is given by the amplitude on that i-th site of the electronic Bloch wave function corresponding to the initial state with energy ϵ_k. These amplitudes are simply the coefficients of the eigenstates for the tight-binding Hamiltonians, as is evident from Eqs. (2.2) and (2.4). The exponential factors $\exp(-i\boldsymbol{G} \cdot \boldsymbol{R}_i)$ take into account the in-plane difference in optical paths between electron waves originating on different sites and travelling towards the detector. Hence, in this simplified description, ARPES patterns correspond to a electron-wave version of quantum Young's double-slit experiment [41]. The number of sources equals in this case the number of atoms in the unit cell. Conservation of the in-plane momentum maps directly the ARPES pattern of the photoelectron momentum to the constant-energy cuts through the band structure of our material, while conservation of energy defines the total wavevector $(\boldsymbol{k} + \boldsymbol{G})$ observable in the experiment. The greater the energy of the incoming photons, the more Brillouin zones can be resolved. Finally, we do not model here dynamical effects that lead to energy broadening [18, 22–27] but introduce a Lorentzian $\delta(\cdots) \approx \pi^{-1}\Gamma/[(\cdots)^2 + \Gamma^2]$ in the figures with the parameter Γ representing finite energy broadening.

3.2 Monolayer Graphene

To derive the angular distributions of the ARPES intensity for monolayer graphene, we use the tight-binding Hamiltonian, which can be obtained in a procedure very similar to that presented for bilayer graphene in Chap. 2. Considering only one layer of hexagonally arranged carbon atoms (for example the bottom one in Fig. 2.1), one ends with the same unit vectors, unit cell and Brillouin zone. However, the unit cell now contains only two atoms, A and B, with their positions within the unit cell given as $\boldsymbol{R}_A = -\boldsymbol{d}_1$ and $\boldsymbol{R}_B = \boldsymbol{0}$, respectively. Within the linear approximation, the Hamiltonian of monolayer graphene in the basis $(\phi_{+,A}, \phi_{+,B})^T$ or $(\phi_{-,B}, \phi_{-,A})^T$ depending on the valley, is [42–44]

$$\hat{\boldsymbol{H}}_{11} = \xi v \begin{pmatrix} 0 & \hat{\pi}^{\dagger} \\ \hat{\pi} & 0 \end{pmatrix}. \tag{3.10}$$

From this follow the energy eigenvalue ϵ_p and corresponding eigenstates ψ_p,

$$\epsilon_p = svp, \quad \psi_p = \frac{1}{\sqrt{2}} \begin{pmatrix} e^{-i\frac{\varphi}{2}} \\ \xi s e^{i\frac{\varphi}{2}} \end{pmatrix}, \tag{3.11}$$

where $\varphi = \arctan \frac{p_y}{p_x}$ and $s = \pm 1$ denotes the conduction ($s = 1$) or the valence ($s = -1$) band. Both the conduction and valence bands have linear dispersion. They touch each other exactly in the center of the valley at the energy ϵ_D usually taken as zero of the energy scale. This characteristic feature leads to the K points often being called the Dirac points [45–47]. With the use of Eq. (3.9), the angular distribution of ARPES for monolayer graphene is related to

$$I \propto \frac{1}{2} \left| e^{-i\frac{\varphi}{2}} e^{-i\frac{\xi}{2}\mathbf{G}\cdot\mathbf{d}_1} + \xi s e^{i\frac{\varphi}{2}} e^{i\frac{\xi}{2}\mathbf{G}\cdot\mathbf{d}_1} \right|^2 \tag{3.12}$$

$$= 1 + \xi s \cos\left[\varphi - \xi \frac{2\pi}{3}(m_2 - m_1) \right].$$

The two real numbers, m_1 and m_2, define which Brillouin zone the ARPES spectrum is described for. However, as the vicinity of any valley belongs formally to one of the three neighbouring Brillouin zones (due to the K point being a corner of a hexagon), for further simplicity of our discussion, we choose different unit cell in the reciprocal space, that is, a rhombus centred on the Γ point, as shown in Fig. 2.1b. This choice does not influence any of the formulae in this chapter, yet simplifies the problem of choosing the numbers m_1 and m_2 as the vicinity of any K point is now contained in a single unit cell. For the two representative valleys K_+ and K_-, Eq. (3.12) can now be put in the simple form

$$I \propto \begin{cases} \cos^2\left[\frac{\varphi}{2} - \xi\frac{\pi}{3}(m_2 - m_1) \right] & \text{for } \xi s = 1 \\ \sin^2\left[\frac{\varphi}{2} - \xi\frac{\pi}{3}(m_2 - m_1) \right] & \text{for } \xi s = -1 \end{cases}. \tag{3.13}$$

We see that as the path around the valley is traversed and the angle φ changes, one peak in the intensity is observed, at the angle $\left[\varphi = \xi\frac{2\pi}{3}(m_2 - m_1) \bmod 2\pi \right]$ if $\xi s = 1$ or $\left[\varphi = \pi + \xi\frac{2\pi}{3}(m_2 - m_1) \bmod 2\pi \right]$ if $\xi s = -1$. The ARPES patterns probing states at the same energy ϵ_p differ between the valleys. However, the angular distribution around the valley K_+ for the valence band is the same as the angular distribution around the valley K_- for the conduction band. The same is true for patterns around K_+ for the conduction band and K_- for the valence band.

Numerically calculated ARPES patterns for monolayer graphene within the general tight-binding model, that is the band structure described as

$$\begin{pmatrix} 0 & -\gamma_0 f(\mathbf{k}) \\ -\gamma_0 f^*(\mathbf{k}) & 0 \end{pmatrix} \psi(\mathbf{p}) = \epsilon_p \begin{pmatrix} 1 & s_0 f(\mathbf{k}) \\ s_0 f^*(\mathbf{k}) & 1 \end{pmatrix} \psi(\mathbf{p}), \tag{3.14}$$

are shown in Fig. 3.1. Although due to additional terms in the electronic momentum $\mathbf{p}$ contained in the geometrical factor f, the eigenstate can no longer be written down in a simple form containing the azimuthal angle φ, the above conclusions still hold. As a particular valley is circled round, the ARPES intensity exhibits one peak (intensity is at maximum) and one dip (intensity is zero). Due to those higher terms in $\mathbf{p}$, for energies far from the Dirac point (Fig. 3.1a) the pattern around each valley is trigonally warped. However, for energies close to the Dirac point (Fig. 3.1b–e), the isoenergetic lines are circular and the electron-hole asymmetry is negligible. Thus, the pattern around the valley K_+ (K_-) for the valence band is the same as around the valley K_- (K_+) for the conduction band.

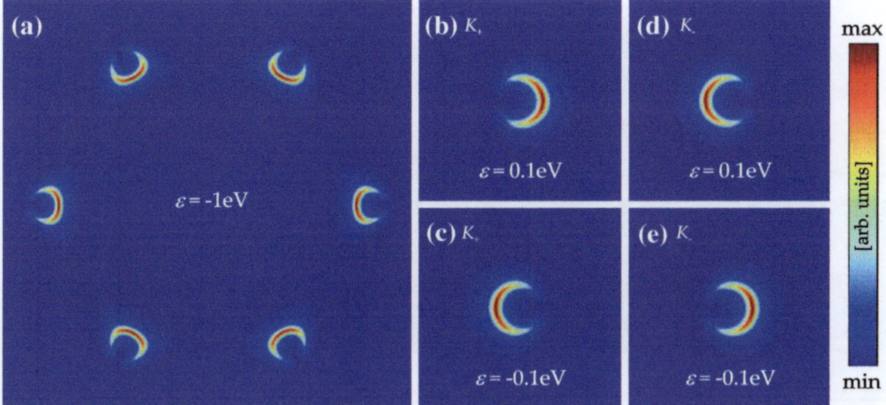

Fig. 3.1 The constant-energy ARPES maps of monolayer graphene obtained within the full tight-binding model, Eq. 3.14, for (**a**) whole Brillouin zone; (**b**) and (**c**) valley K_+; (**d**) and (**e**) valley K_-. Energy corresponding to each map is given with respect to the Dirac point. The length of the side of the map in the reciprocal space is $\frac{10\pi}{3a}$ and $\frac{8\pi}{75a}$ for (**a**) and (**b**)–(**e**), respectively. Numerical values of the parameters used: $\gamma_0 = 3$ eV, $s_0 = 0.129$, $a = 2.46$ Å

The peculiar behaviour of the ARPES intensity as a function of the azimuthal angle φ around the valley K_ξ has been noticed before for bulk graphite [9]. As shown, its origins lie purely in the hexagonal symmetry of graphene. This symmetry gives rise to a specific phase relation between components of the electron wave on the two sublattices and the electron's momentum. In recent literature, this is often referred to as *chirality* of electrons in graphene [45–48]. The angle-resolved photoemission provides thus a direct observation of this phenomenon.

3.3 Bilayer Graphene

As explained in the previous section, the main features in the angular distribution of the ARPES intensity for graphene have been in the past observed for graphite. As graphene and graphite are conceptually the extreme cases of hexagonally layered carbon system, one might expect no significant differences in the ARPES spectra for systems in between, that is also for bilayer graphene. However, the low-energy electronic spectrum undergoes a drastic change from linear to quadratic dispersion when two graphene layers are coupled together to form a bilayer. The question is therefore, whether this change leads to new features in the angular distribution of the ARPES intensity.

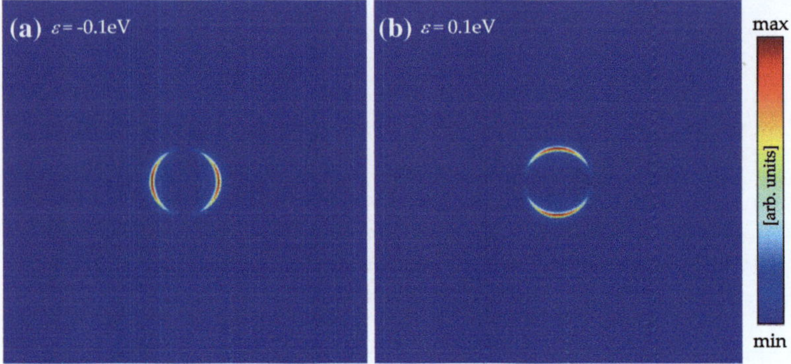

Fig. 3.2 The constant-energy ARPES maps of bilayer graphene obtained within the low-energy two-band approximation, Eq. 3.14, for energies (**a**) $\epsilon = -0.1\,$eV, (**b**) $\epsilon = 0.1\,$eV with respect to the neutrality point. The length of the side of the map in the reciprocal space is $\frac{4\pi}{15a}$, numerical values of the parameters used: $v = 0.97\,$m/s, $\gamma_1 = 0.4\,$eV

3.3.1 Low-Energy Spectrum: Contribution of the Two Degenerate Bands Only

As we are concerned with the low-energy electronic properties, we start with the effective two-band approximation, Eq. (2.12). For simplicity of the arguments that follow, we neglect the v_3 terms and write the eigenproblem

$$-\frac{v^2}{\gamma_1}\begin{pmatrix} 0 & \left(\hat{\pi}^{\dagger}\right)^2 \\ \hat{\pi}^2 & 0 \end{pmatrix}\psi = \epsilon\psi. \tag{3.15}$$

We write the resulting spectrum and electronic eigenstates as follows:

$$\epsilon_{\boldsymbol{p}} = s\frac{v^2}{\gamma_1}p^2; \quad \psi = \frac{1}{\sqrt{2}}\begin{pmatrix} e^{-i\varphi} \\ -se^{i\varphi} \end{pmatrix}, \tag{3.16}$$

where s, as before, swaps between the conduction ($s = 1$) and valence ($s = -1$) bands and $\varphi = \arctan\frac{p_y}{p_x}$. The eigenstates above are very similar to those in the case of the monolayer, Eq. (3.11), although the prefactor $\frac{1}{2}$ in front of the angle φ and valley index ξ are missing. From Eq. (3.9), the ARPES intensity is

$$I \propto \begin{cases} \cos^2\left[\varphi + \xi\frac{2\pi}{3}\left(m_2 - m_1\right)\right] & \text{for } s = -1 \\ \sin^2\left[\varphi + \xi\frac{2\pi}{3}\left(m_2 - m_1\right)\right] & \text{for } s = 1 \end{cases}. \tag{3.17}$$

Comparison between (3.17) and (3.13 shows that for bilayer graphene the angular distribution of the ARPES intensity around a valley should depends on twice the characteristic angle obtained for monolayer, $\left[\frac{\varphi}{2} - \frac{\pi}{3}\left(m_2 - m_1\right)\right]$. Therefore, two symmetric peaks and two dips in the intensity are expected at low energies as the path around the valley is traversed. That is shown in the Fig. 3.2. No difference (except possibly trigonal warping effects) should occur when changing the valley from K_+ to K_-. Also, a rotation by $\pi/2$ should occur when swapping the band.

3.3.2 Contribution from the Split Bands

In the previous section we described the ARPES spectra in the low-energy limit. At the same time, the ARPES pattern should evolve so that at higher energies it resembles that of monolayer graphene or graphite, Fig. 3.1a, with one peak in the intensity. Such a regime can not be described with the low energy approximation. We move therefore to the four-band Hamiltonian, which not only enables us better comparison to experimental data, but also, as it turns out, adds to the low-energy description from the previous section [40]. Again, for the sake of the argument, we start with the minimum model describing all four bands around the valley K_ξ, that is,

$$\begin{pmatrix} 0 & 0 & 0 & \xi v \pi^\dagger \\ 0 & 0 & \xi v \pi & 0 \\ 0 & \xi v \pi^\dagger & 0 & \gamma_1 \\ \xi v \pi & 0 & \gamma_1 & 0 \end{pmatrix} \psi = \epsilon \psi. \tag{3.18}$$

Once again, we require the knowledge of the eigenstates and the energies,

$$\epsilon_{sb} = s\frac{1}{2}\left(\sqrt{\gamma_1^2 + 4v^2 p^2} + b\gamma_1\right), \quad \psi_{sb} = \frac{1}{\sqrt{2}\sqrt{1 + \left(\frac{|\epsilon_{sb}|}{vp}\right)^2}} \begin{pmatrix} e^{-i\varphi} \\ sbe^{i\varphi} \\ \xi b \frac{|\epsilon_{sb}|}{vp} \\ \xi s \frac{|\epsilon_{sb}|}{vp} \end{pmatrix},$$
$$\tag{3.19}$$

where $b = \pm 1$ distinguishes between the split and the low-energy bands. Note that what b (1 or -1) corresponds to which set of bands depends on the sign of γ_1. This does not matter for the band structure (does not change the form of the eigenvalues, only reorders them), but does influence the form of the eigenstate corresponding to each band. The angular distribution of the ARPES intensity is proportional to

$$I \propto \left(1 + sb\cos\left[2\varphi + \xi\frac{4\pi}{3}(m_2 - m_1)\right]\right.$$
$$\left. + 2\delta_{sb}\left(\frac{|\epsilon_{bs}|}{vp}\right)\left\{\frac{|\epsilon_{bs}|}{vp} + (b+s)\cos\left[\varphi + \xi\frac{2\pi}{3}(m_2 - m_1)\right]\right\}\right). \tag{3.20}$$

The first two terms correspond to the low energy limit discussed in the previous section. The last term is then a correction, which vanishes for $sb = -1$. In this case one is left with the pattern of two symmetric peaks, as in Fig. 3.2. However, this happens for only one of the two low-energy bands (and one high-energy band). For the other, $sb = 1$, the correction does not vanish and contributes towards the intensity. To estimate its importance, we consider valleys $K_{\pm} = (\pm\frac{4\pi}{3a}, 0)$ in the first Brillouin zone ($m_1, m_2 = 0$). Also, for the low energy bands, $|\epsilon| \approx \frac{v^2}{\gamma_1}p^2$. Then,

$$I \propto \left(1 + \cos(2\varphi) + 4s\sqrt{\frac{|\epsilon_{sb}|}{\gamma_1}}\cos\varphi + 2\frac{|\epsilon_{sb}|}{\gamma_1} \right).$$

This function has two maxima for $\varphi \in (0, 2\pi)$, at $\varphi = 0$ and $\varphi = \pi$. The ratio of their intensities is

$$\frac{I(\varphi = 0)}{I(\varphi = \pi)} = \left(\frac{1 + s\sqrt{\frac{|\epsilon|}{\gamma_1}}}{1 - s\sqrt{\frac{|\epsilon|}{\gamma_1}}} \right)^2 .$$

For the energy $|\epsilon_{sb}| = 0.025\,\text{eV}$ from the neutrality point, the above ratio yields $\frac{25}{9}$ for $s = 1$ or $\frac{9}{25}$ for $s = -1$. Hence, one of the peaks is more than twice higher than the other. That strong asymmetry is obtained for energies of the order of $\frac{\gamma_1}{16}$, which are often considered to be in all instances well described by the two-band approximation.

To summarise, according to Eq. (3.20), we expect the symmetric two-peak pattern to appear for two of the bands. For $\gamma_1 > 0$, these are the valence split band ($s = -1$, $b = 1$) and the conduction low-energy band ($s = 1$, $b = -1$). For the other two bands, we expect one of the peaks to go darker as we increase the distance from the neutrality point. From the band structure considerations, Eq. (2.9), we also expect the pattern to be strongly trigonally warped at low energies due to the coupling γ_3. This low-energy warping does not occur in the monolayer case.

The constant-energy maps of the ARPES intensity, calculated within the full four-band model, Eq. (2.4) and (2.6), are presented in Fig. 3.3 for the whole Brillouin zone, (a), and valley K_+, (b)–(g). As anticipated, for the energy far from the neutrality point, Fig. 3.3a, the ARPES spectrum look similar to that of the monolayer, 3.1a and graphite [9, 39]. Figures 3.3b–g show the evolution of the ARPES pattern as the energy changes from 0.5 to $-0.5\,\text{eV}$. At energies greater than the interlayer coupling, $\epsilon > \gamma_1$, (b) and (c), there are two ring-like patterns, each corresponding to photoemission from states in two bands, whereas, for low-energies, $\epsilon < \gamma_1$, (d)–(g), there is a single ring corresponding to emission from the degenerate band only. The dot in the center of Fig. 3.3d corresponds to the photoemission from the bottom of the split band due to finite energy width Γ. The disappearance of this dot provides an estimate for the magnitude of the parameter γ_1 [31]. It is not visible on Fig. 3.3e, because for this band $sb = -1$. As mentioned before, the sign of γ_1 determines for which set of bands the pattern of two equally bright peaks should be observed. In agreement with the conclusions from Eq. 3.20, in the Fig. 3.3b–g these symmetric peaks are observed for the low-energy conduction and high-energy valence bands as γ_1 was taken to be positive. Negative sign of γ_1 would lead to equally bright

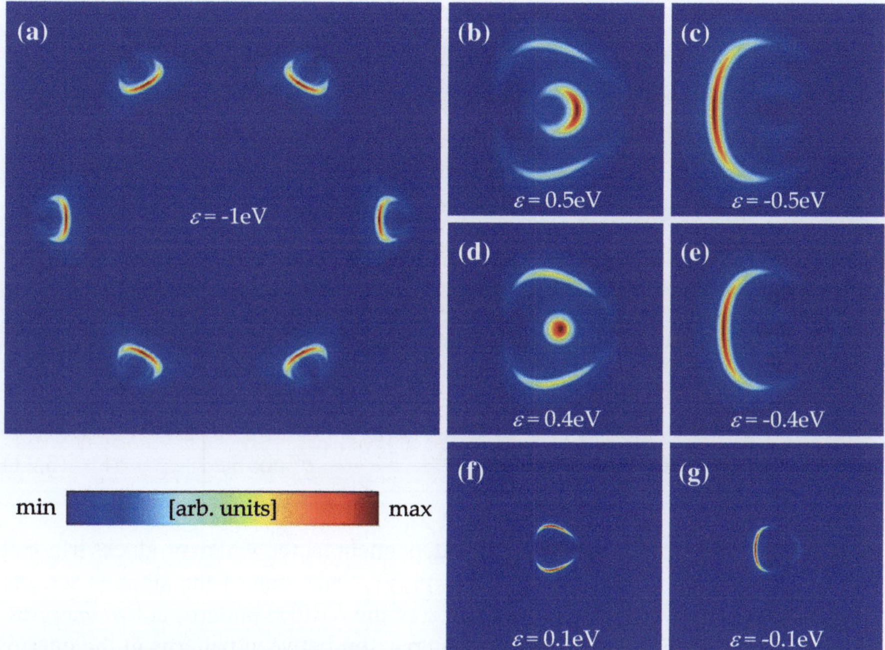

Fig. 3.3 The constant-energy ARPES maps of bilayer graphene obtained within the four-band tight-binding model, Eq. (2.4) and (2.6). All energies are given with respect to the neutrality point. The length of the side of the map in the reciprocal space is (**a**) $\frac{16\pi}{5a}$, (**b**)–(**g**) $\frac{4\pi}{15a}$. Numerical values of the parameters used: $\gamma_0 = 3\,\text{eV}$, $a = 2.46\,\text{Å}$ (resulting $v \approx 0.97\,\text{m/s}$), $\gamma_1 = 0.4\,\text{eV}$, $\gamma_3 = 0.2\,\text{eV}$, $s_0 = 0.129$, and the energy width Γ is one sixth of the corresponding energy

peaks appearing in the high-energy conduction and low-energy valence bands. Hence, probing the electron wave function via ARPES may reveal not only the magnitude of the coupling but also its sign. We emphasize that this sign is not important when the tight-binding model is used only to obtain the band structure and comparisons to experimentally obtained electronic dispersions relate only to the magnitude of the band splittings which are always positive. Also, notice the second peak evolving in the pattern originating from the low-energy valence band at $\varphi = 0$ as the energy shifts closer to the neutrality point. It is not visible at all at the energy $\epsilon = -0.5\,\text{eV}$, but is quite clear for $\epsilon = -0.1\,\text{eV}$, where, in agreement to the above discussion, it is less than half of the dominant peak around $\varphi = \pi$.

As mentioned above, the sign of the γ_1 coupling may be extracted from the ARPES spectra. We now show that knowledge of this sign allows for the determination of the sign of the trigonal warping parameter, γ_3. Let us recall the low energy dispersion around the valley K_ξ as described by the two-band approximation, Eq. (2.13),

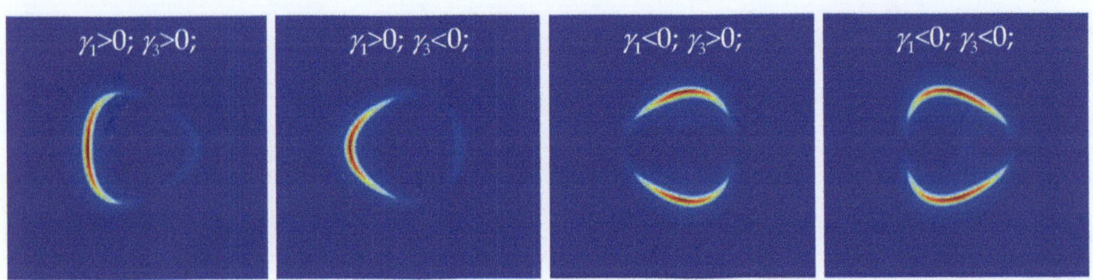

Fig. 3.4 The influence of the signs of γ_1 and γ_3 on the ARPES spectra for the energy $\epsilon = -0.1\,\text{eV}$ with respect to the neutrality point. The length of the side of the map in the reciprocal space is $\frac{4\pi}{30a}$. Numerical values of the parameters used: $\gamma_0 = 3\,\text{eV}$, $a = 2.46\,\text{Å}$ (resulting $v \approx 0.97\,\text{m/s}$), $|\gamma_1| = 0.4\,\text{eV}$, $|\gamma_3| = 0.2\,\text{eV}$, $s_0 = 0.129$, and the energy width $\Gamma = 0.1/6\,\text{eV}$

$$\epsilon_\xi = s\left[\left(\frac{v^2}{\gamma_1}\right)^2 p^4 + v_3^2 p^2 - \frac{2\xi v^2 v_3}{\gamma_1} p^3 \cos 3\varphi\right]^{\frac{1}{2}}. \tag{3.21}$$

This expression illustrates that the angular dependent factor, which produces trigonal warping, depends on the sign of the ratio γ_3/γ_1. Once one of the signs is set, the other follows from investigation of the shape of the ARPES patterns at low energies. This is presented in Fig. 3.4, where the comparison between patterns at the energy $-0.5\,\text{eV}$ for different signs of γ_1 and γ_3 is shown. For $\text{sgn}\gamma_1 = \text{sgn}\gamma_3$, the trigonal warping due to γ_3 deforms the isoenergetic lines in the same fashion as the high-energy trigonal warping due to higher than linear terms in the electronic momentum in the factor $f(\mathbf{k})$ [Fig. 3.4a, d; compare to Fig. 3.3a]. However, for $\text{sgn}\gamma_1 = -\text{sgn}\gamma_3$, the low-energy warping counteracts the effects of the high-energy warping [Fig. 3.4b, c]. Once the sign of γ_1 has been established, the direction of the low-energy warping along the p_x axis [compare Fig. 3.4a, b or c, d] determines the sign of γ_3.

3.3.3 Influence of the Symmetry-Breaking Parameters on the ARPES Spectra

The angular distributions of the ARPES intensity are also sensitive to the symmetry-breaking parameters u, Δ and Δ_{AB}. There are two reasons for that: (I) all of these parameters, as shown in Sect. 2.2.3, modify the band structure, thus changing the shape of the isoenergetic lines in the reciprocal space probed with ARPES; (II) these parameters also influence the amplitude C_i of the electron Bloch wave on the atomic site i. The influence of the on-site asymmetries on the ARPES spectra is shown on the example of the interlayer asymmetry u in Fig. 3.5. Four spectra around the K_+ valley for different values of the asymmetry are presented, as well as the corresponding low-energy band structures (beneath each ARPES spectrum). The two-peak pattern is quite robust against the opening of a gap. Only when the top/bottom of the gap is near the probed energy, the pattern is distorted (last column in Fig. 3.5). The opening

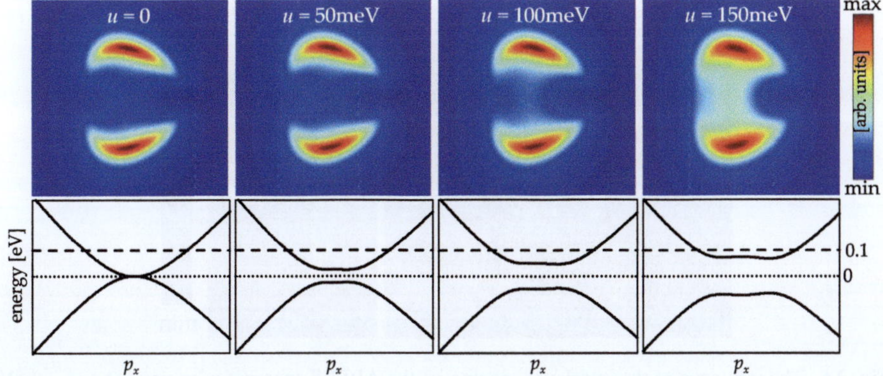

Fig. 3.5 The influence of the interlayer asymmetry u on the ARPES spectra for energy $\epsilon = 0.1\,\text{eV}$ with respect to the neutrality point. The length of the side of the map in the reciprocal space is $\frac{8\pi}{75a}$. Shown below the spectra are corresponding electronic dispersions along the P_x axis; the *dotted lines* show the position of the neutrality point (the middle of the gap) and the *dashed lines* show the energy the spectra are resolved for. Numerical values of the parameters used: $\gamma_0 = 3\,\text{eV}$, $a = 2.46\,\text{Å}$ (resulting $v \approx 0.97\,\text{m/s}$), $\gamma_1 = 0.4\,\text{eV}$, $\gamma_3 = 0.2\,\text{eV}$, $s_0 = 0.129$, and the energy width $\Gamma = 0.1/3\,\text{eV}$

of the gap in the electronic spectrum due to the interlayer asymmetry u has already been successfully observed with ARPES [31].

3.3.4 Interference Due to a Finite Interlayer Distance

Up to this point, we treated bilayer graphene as purely two-dimensional and disregarded the existence of the interlayer spacing between layers, c_0. Clearly, c_0 is going to influence the ARPES spectra via the phase factors describing the difference in the optical paths for photoelectron waves originating on different atomic sites, as shown in Eq. (3.8). The out-of-plane component of the momentum is

$$p_e^{\perp} = \sqrt{2m\,(\omega - W + \epsilon_k) - \hbar^2(\boldsymbol{k} + \boldsymbol{G})^2}, \qquad (3.22)$$

and we assume that only the photoelectrons with $p_e^{\perp} > 0$ are detected. In order to calculate the ARPES patterns, we now need to set the values of the energy of the incoming photons ω and the material constant, work function W. The intensity distribution now depends on ω.

Examples of the ARPES spectra for different energies of the incoming radiation have been shown in Fig. 3.6. At high-energies (Fig. 3.6a) the pattern does not qualitatively change. However, at low energies (Fig. 3.6b–c), when the contribution due to the states in the split bands is no longer present, the dependence on ω manifests itself through a rotation of the pattern around the valley. These slightly rotated patterns resemble experimental constant-energy maps of ARPES intensity shown in

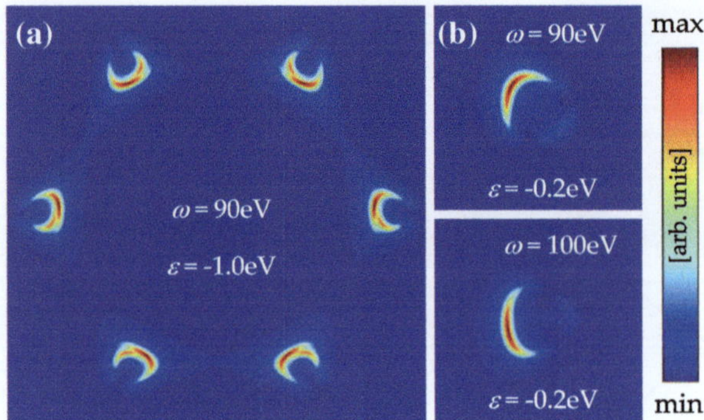

Fig. 3.6 The influence of the interlayer spacing on the ARPES spectra for energy (**a**) $\epsilon = -1$ eV, (**b**)–(**c**) $\epsilon = -0.2$ eV with respect to the neutrality point. The length of the side of the map in the reciprocal space is (**a**) $\frac{16\pi}{5a}$, (**b**)–(**c**) $\frac{4\pi}{15a}$. Numerical values of the parameters used: $\gamma_0 = 3$ eV, $a = 2.46$ Å (resulting $v \approx 0.97$ m/s), $\gamma_1 = 0.4$ eV, $\gamma_3 = 0.2$ eV, $s_0 = 0.129$, $c_0 = 3.4$ Å, $W = 5$ eV and the energy width (**a**) $\Gamma = 0.2$ eV, (**b**)–(**c**) $\Gamma = 0.04$ eV

the online material supporting work published in Ref. [31]. The additional rotation caused by the phase factor associated with the interlayer distance c_0 makes extracting the band structure parameters, and especially their signs, from the spectra much more difficult. However, with detailed comparison with experiment and calibration of the energy scale, it may still be possible. Then, an energy can be chosen for the incoming photons so that $\frac{p_e^{\perp} c_0}{2} \approx 0$ for photoelectrons originating from the states close to the centre of the valley.

References

1. H. Hertz, Ueber einen Einfluss des ultravioletten Lichtes auf die electrische Entladung. Ann. Phys. **267**, 983 (1887)
2. A. Einstein, Über einen die Erzeugung und Verwandlung des Lichtes betreffenden heuristischen Gesichtspunkt. Ann. Phys. **322**, 132 (1905)
3. P.M. Williams, The direct evaluation of electronic band structures of layered solids using angle-resolved photoemission. Il Nuovo Cimento **38B**, 216 (1977)
4. I.T. McGovern, W. Eberhardt, E.W. Plummer, J.E. Fischer, The bandstructures of graphite and graphite intercalation compounds as determined by angle resolved photoemission using synchrotron radiation. Physica **99B**, 415 (1980)
5. W. Eberhardt, I.T. McGovern, E.W. Plummer, J.E. Fisher, Charge-transfer and non-rigid-band effects in the graphite compound LiC_6. Phys. Rev. Lett. **44**, 200 (1980)
6. A.R. Law, J.J. Barry, H.P. Hughes, Angle-resolved photoemission and secondary electron emission from single-crystal graphite. Phys. Rev. B **28**, 5332 (1983)
7. T. Takahashi, H. Tokailin, S. Sagawa, Electronic band structure of graphite studied by highly angle-resolved ultraviolet photoelectron spectroscopy. Solid State Commun. **52**, 765 (1984)
8. A. Santoni, L.J. Terminello, F.J. Himpsel, T. Takahashi, Mapping the Fermi surface of graphite with a display-type photoelectron spectrometer. Appl. Phys. A **52**, 299 (1991)

9. E.L. Shirley, L.J. Terminello, A. Santoni, F.J. Himpsel, Brillouin-zone-selection effects in graphite photoelectron angular distributions. Phys. Rev. B **51**, 13614 (1995)

10. E. Rollings, G.-H. Gweon, S.Y. Zhou, B.S. Mun, J.L. McChesney, B.S. Hussain, A.V. Federov, P.N. First, W.A. de Heer, A. Lanzara, Synthesis and characterization of atomically thin graphite films on a silicon carbide substrate. J. Phys. Chem. Solids **67**, 2172 (2006)

11. M. Sprinkle, D. Siegel, Y. Hu, J. Hicks, A. Tejede, A. Taleb-Ibrahimi, P. Le Fèvre, F. Bertran, S. Vizzini, H. Enriquez, S. Chiang, P. Soukiassian, C. Berger, W.A. de Heer, A. Lanzara, E.H. Conrad, First direct observation of a nearly ideal graphene band structure. Phys. Rev. Lett. **103**, 226803 (2009)

12. Y.S. Dedkov, M. Fonin, C. Laubschat, A possible source of spin-polarized electrons: the inert graphene/Ni(111) system. Appl. Phys. Lett. **92**, 052506 (2008)

13. Y.S. Dedkov, M. Fonin, U. Rüdiger, C. Laubschat, Rashba effect in the graphene/Ni(111) system. Phys. Rev. Lett. **100**, 107602 (2008)

14. A. Grüneis, D.V. Vyalikh, Tunable hybridization between electronic states of graphene and a metal surface. Phys. Rev. B **77**, 193401 (2008)

15. A. Grüneis, K. Kummer, D.V. Vyalikh, Dynamics of graphene growth on a metal surface: a time-dependent photoemission study. New J. Phys. **11**, 073050 (2009)

16. I. Pletikosić, M. Kralj, P. Pervan, R. Brako, J. Coraux, A.T. N'Diaye, C. Busse, T. Michely, Dirac cones and minigaps for graphene on Ir(111). Phys. Rev. Lett. **102**, 056808 (2009)

17. C. Enderlein, Y.S. Kim, A. Bostwick, E. Rotenberg, K. Horn, The formation of an energy gap in graphene on ruthenium by controlling the interface. New J. Phys. **12**, 033014 (2010)

18. A. Bostwick, T. Ohta, T. Seyller, K. Horn, E. Rotenberg, Quasiparticle dynamics in graphene. Nat. Phys. **3**, 36 (2007)

19. A. Bostwick, T. Ohta, J.L. McChesney, T. Seyller, K. Horn, E. Rotenberg, Renormalization of graphene bands by many-body interactions. Solid State Commun. **143**, 63 (2007)

20. A. Bostwick, T. Ohta, J.L. McChesney, T. Seyller, K. Horn, E. Rotenberg, Band structure and many body effects in graphene. Eur. Phys. J. **148**, 5 (2007)

21. S.Y. Zhou, G.-H. Gweon, A.V. Fedorov, P.N. First, W.A. de Heer, D.-H. Lee, F. Guinea, A.H. Castro Neto, A. Lanzara, Substrate-induced bandgap opening in epitaxial graphene. Nat. Mater. **6**, 770 (2007)

22. C.-H. Park, F. Giustino, M.L. Cohen, S.G. Louie, Velocity renormalization and carrier lifetime in graphene from the electron-phonon interaction. Phys. Rev. Lett. **99**, 086804 (2007)

23. M. Calandra, F. Mauri, Electron-phonon coupling and electron self-energy in electron-doped graphene: calculation of angular-resolved photoemission spectra. Phys. Rev. B **76**, 205411 (2007)

24. W.-K. Tse, S. Das Sarma, Phonon-induced many-body renormalization of the electronic properties of graphene. Phys. Rev. Lett. **99**, 236802 (2007)

25. M. Polini, R. Asgari, G. Borghi, Y. Barlas, T. Pereg-Barnea, A.H. MacDonald, Plasmons and the spectral function of graphene, Phys. Rev. B **77**, 081411(R) (2008)

26. P.E. Trevisanutto, C. Giorgetti, L. Reining, M. Ladisa, V. Olevano, Ab initio GW many-body effects in graphene. Phys. Rev. Lett. **101**, 226405 (2008)

27. C.-H. Park, F. Giustino, C.D. Spataru, M.L. Cohen, S.G. Louie, Angle-resolved photoemission spectra of graphene from first-principles calculations. Nano Lett. **9**, 4234 (2009)

28. A. Bostwick, T. Ohta, J.L. McChesney, K.V. Emtsev, T. Seyller, K. Horn, E. Rotenberg, Symmetry breaking in few layer graphene films. New J. Phys. **9**, 385 (2007)

29. E. Rotenberg, A. Bostwick, T. Ohta, J.L. McChesney, T. Seyller, K. Horn, Origin of the energy bandgap in epitaxial graphene, Nat. Mat. **7**, 258 (2007) [comment on Zhou et al., [21]]

30. S.Y. Zhou, G.-H. Gweon, A.V. Fedorov, P.N. First, W.A. de Heer, D.-H. Lee, F. Guinea, A.H. Castro Neto, A. Lanzara, Nat. Mat. **7**, 259 (2007), Authors' response [to the comment by Rotenberg et al., [29]]

31. T. Ohta, A. Bostwick, T. Seyller, K. Horn, E. Rotenberg, Controlling the electronic structure of bilayer graphene. Science **313**, 951 (2006)

32. S.Y. Zhou, D.A. Siegel, A.V. Fedorov, A. Lanzara, Metal to insulator transition in epitaxial graphene induced by molecular doping. Phys. Rev. Lett. **101**, 086402 (2008)

33. C. Coletti, C. Riedl, D.S. Lee, B. Krauss, L. Patthey, K. von Klitzing, J.H. Smet, U. Starke, Charge neutrality and band-gap tuning of epitaxial graphene on SiC by molecular doping. Phys. Rev. B **81**, 235401 (2010)
34. T. Ohta, A. Bostwick, J.L. McChesney, T. Seyller, K. Horn, E. Rotenberg, Interlayer interaction and electronic screening in multilayer graphene investigated with angle-resolved photoemission spectroscopy. Phys. Rev. Lett. **98**, 206802 (2007)
35. A. Bostwick, K.V. Emtsev, K. Horn, E. Huwald, L. Ley, J.L. McChesney, T. Ohta, J. Riley, E. Rotenberg, F. Speck, T. Seyller, Photoemission studies of graphene on SiC: growth, interface, and electronic structure. Adv. Solid State Phys. **47**, 159 (2008)
36. A. Bostwick, J.L. McChesney, T. Ohta, E. Rotenberg, T. Seyller, K. Horn, Experimental studies of the electronic structure of graphene. Prog. Surf. Sci. **84**, 380 (2009)
37. F.J. Himpsel, Angle-resolved measurements of the photoemission of electrons in the study of solids. Adv. Phys. **32**, 1 (1983)
38. A. Damascelli, Probing the electronic structure of complex systems by ARPES. Phys. Scr. **T109**, 61 (2004)
39. S.Y. Zhou, G.-H. Gweon, J. Graf, A.V. Fedorov, C.D. Spataru, R.D. Diehl, Y. Kopelevich, D.-H. Lee, S.G. Louie, A. Lanzara, First direct observation of Dirac fermions in graphite. Nat. Phys. **2**, 595 (2006)
40. M. Mucha-Kruczyński, O. Tsyplyatyev, A. Grishin, E. McCann, V.I. Fal'ko, A. Bostwick, E. Rotenberg, Characterization of graphene through anisotropy of constant-energy maps in angle-resolved photoemission. Phys. Rev. B **77**, 195403 (2008)
41. R.P. Feynman, *The Feynman Lectures on Physics*, vol III (Addison-Wesley, Reading, Massachusetts, 1963)
42. P.R. Wallace, The band theory of graphite. Phys. Rev. **71**, 622 (1947)
43. D.P. DiVincenzo, E.J. Mele, Self-consistent effective-mass theory for intralayer screening in graphite intercalation compounds. Phys. Rev. B **29**, 1685 (1984)
44. C. Bena, G. Montambaux, Remarks on the tight-binding model of graphene. New J. Phys. **11**, 095003 (2009)
45. A.K. Geim, K.S. Novoselov, The rise of graphene. Nat. Mat. **6**, 183 (2007)
46. M.I. Katsnelson, K.S. Novoselov, Graphene: new bridge between condensed matter physics and quantum electrodynamics. Solid State Commun. **143**, 3 (2007)
47. A.H. Castro Neto, F. Guinea, N.M.R. Peres, K.S. Novoselov, A.K. Geim, The electronic properties of graphene. Rev. Mod. Phys. **81**, 109 (2009)
48. M.I. Katsnelson, K.S. Novoselov, A.K. Geim, Chiral tunnelling and the Klein paradox in graphene. Nat. Phys. **2**, 620 (2006)

Chapter 4
Magneto-Optical Spectroscopy

The behaviour of electrons in (quasi-)two-dimensional systems in external magnetic fields is a fascinating area of physics. Classically, the Lorentz force caused by the magnetic field curves the trajectory of a charged particle. If such a particle is constrained to move only in one plane, in a strong enough field perpendicular to that plane, the trajectory of the particle becomes a closed orbit. However, quantum mechanically, due to wave nature of matter, only some of the orbits are stable. For a two-dimensional solid in low temperatures, this results in the electronic band structure turning into a discrete spectrum of Landau levels (LLs) [1]. The number of states per unit area in each Landau level (LL) (degeneracy of the LL) is equal to those originally within the range of one cyclotron energy $\omega_c = \frac{eB}{m}$ (where m is the effective mass of the electron) in the two-dimensional density of states, that is $\frac{eB}{h}$ per LL. The number of the LLs filled with electrons is described by the filling factor $\nu = n\frac{h}{eB}$ (with $\nu = 0$ corresponding to the neutral system). As ν is varied, for example by changing the applied magnetic field, the Landau level crossing the Fermi energy is filled or emptied of electrons. Repetitive crossing of the Fermi energy by Landau levels leads, for example, to oscillations in the conductivity measured as a function of the magnetic field (Shubnikov-de Haas effect) [2]. Similar in origin is the appearance of discrete steps in the Hall conductivity σ_{xy} in the integer quantum Hall effect (QHE) [3, 4]. In fact, it is the observation of the unusual sequencing of these steps for monolayer and bilayer graphene [5–8] that fuelled most of the initial interest in the Landau level structure of graphene systems. For monolayer, the sequence is shifted with respect to the QHE sequence of a 2DEG, so that $\sigma_{xy} = \pm g\frac{e^2}{h}(N + \frac{1}{2})$, where N is the Landau level index and g is the level degeneracy (in graphene materials it is 4 due to double valley and double spin degeneracy). For bilayer, the steps appear at $\sigma_{xy} = \pm g\frac{e^2}{h}N$. However, the plateau at $\sigma_{xy} = 0$ is missing. Both situations can be easily understood with the help of the Landau level structure. In particular, as the electronic structure of both systems is gapless (disregarding for a moment any on-site asymmetries), an addiditional, unusual Landau level is present at the energy $\epsilon = 0$, sharing states between electrons and holes. Hence, both for monolayer and bilayer, no $\nu = 0$ plateau exists in symmetric structures. For monolayer,

M. Mucha-Kruczyński, *Theory of Bilayer Graphene Spectroscopy*,
Springer Theses, DOI: 10.1007/978-3-642-30936-6_4,
© Springer-Verlag Berlin Heidelberg 2013

integer filling factors follow then the sequence $\nu = \pm 2, \pm 6, \pm 10, \ldots$, leading to $\sigma_{xy} = \pm \frac{4e^2}{h}(N + \frac{1}{2})$. For bilayer, as we will show, the $\epsilon = 0$ level contains twice as much electron states as other LLs. Thus, at least in strong magnetic fields, the plateau sequence $\nu = \pm 4, \pm 8, \pm 12, \ldots$, results in $\sigma_{xy} = \pm \frac{4e^2}{h} N$.

One of the ways to study the Landau level spectrum of a (quasi-)two-dimensional semiconductor heterostructure is to examine its optical absorption spectrum in an external magnetic field, usually perpendicular to the plane of the sample. This method, called simply magneto-optical (absorption) spectroscopy [9], has been extensively applied to graphene systems [10–19], mainly in relation to the unusual $\sqrt{B}$ -dependence of the Landau level energy on the magnetic field and the physics of the zero-energy Landau level at very high fields. A review of the magneto-optical absorption spectroscopy of graphene systems is given in the broader context of optical properties of graphene multilayers in Ref. [20]. In monolayer graphene specifically, these studies confirmed the unequally spaced Landau level spectrum, arising from the linear electronic dispersion in the absence of the magnetic field, and the scaling of the Landau level energies as $\sqrt{B}$. Some deviations from the predictions of the tight-binding model with regards to the transition energies suggested contribution of many-particle interactions to the picture [11]. This is different from the two-dimensional systems with parabolic dispersion, where electron-electron interactions have no impact on the Landau level transition energies in magneto-optical experiments ('Kohn's theorem') [21]. For bilayer graphene, with the electronic dispersion quadratic at energies $\epsilon \ll \gamma_1$, the non-interacting theory predicts at low energies a linear scaling of the Landau level energy with the strength of the magnetic field [22]. At higher energies, $\epsilon \approx \gamma_1$, the Landau level energy should follow a $\sqrt{B}$-dependence. Thus, the Landau level spectrum of bilayer graphene should change from that characteristic of a parabolic dispersion to that of a linear dispersion. This has been observed experimentally by Henriksen et al. [14]. However, the changeover to a $\sqrt{B}$ behaviour occured at lower energies, and more suddenly, than expected. In fact, for some filling factors, a better fit was achieved when fitted to monolayer dispersion rather than bilayer one. Again, the many-body interactions were suggested as responsible for this departure from the predictions of the single particle theory.

In this chapter, we discuss the magneto-optical absorption spectroscopy of bilayer graphene and test the limits of the tight-binding approach as applied to the experimental situation of Henriksen and co-workers. In particular, we investigate the importance of the interlayer asymmetry in that experiment. This chapter is divided into three parts. In the first one, Sect. 4.1, we describe the Landau level structure of bilayer graphene using both the two-band and the four-band models. In this, we follow an approach applied to graphite [23–25] and routinely used in the case of graphene systems [22, 26, 27]. We then derive in Sect. 4.2 selection rules for the optical absorption in magnetic field. We describe the optical strengths of transitions between any of the π bands and include into the model presence of the physically most relevant asymmetries. We also show the resulting magneto-optical spectra. In the last part, Sect. 4.3, we concentrate on the role of interlayer asymmetry in the abovementioned experiment of Henriksen et al., results of which have been initially fitted to the predictions of the 'neutral bilayer' model (tight-binding model with the interlayer asymmetry equal to zero). We show that the experimental setup may have caused significant charge asymmetries between the layers and thus rendered the 'neutral bilayer' model unapplicable. We demonstrate that self-consistently obtained values of the interlayer asymmetry in the presence of the magnetic field help to explain some of the discrepancies between experimental results and the theory used to interpret them. Some of the results contained in this chapter have been published in Refs. [28, 29].

4.1 Bilayer Graphene in an External Magnetic Field

4.1.1 Landau Levels in the Two-Band Model

In the following, we will work in the Landau gauge, $A = (0, Bx)$, and the resulting magnetic field

$$B = \nabla \times A = (0, 0, B)$$

perpendicular to the sample. Let us consider a Landau function $\psi_{n,q} = \frac{1}{\sqrt{L_y}} e^{iqy} \phi_n (x - q\lambda_B^2)$, where L_y is the length of the system in the y direction and $\phi_n(x)$ is the eigenfunction of the quantum harmonic oscillator. We then observe, that operators $\hat{\pi} = p_x + ip_y$ and $\hat{\pi}^\dagger = p_x - ip_y$, with the the electronic momentum now containing the electromagnetic vector potential, $p = -i\hbar\nabla - eA$, coincide with the raising and lowering operators in the space of functions $\psi_{n,q}$. That is,[1]

$$\hat{\pi}^\dagger \psi_n = i\frac{\hbar}{\lambda_B}\sqrt{2(n+1)}\psi_{n+1}, \tag{4.1a}$$

$$\hat{\pi} \psi_n = -i\frac{\hbar}{\lambda_B}\sqrt{2n}\psi_{n-1}, \tag{4.1b}$$

where the magnetic length $\lambda_B = \sqrt{\frac{\hbar}{eB}}$.

Looking now at the leading term in the two-band approximation,[2]

$$\hat{H}_{\text{eff}} = -\frac{v^2}{\gamma_1} \begin{pmatrix} 0 & \left(\hat{\pi}^\dagger\right)^2 \\ \hat{\pi}^2 & 0 \end{pmatrix}, \tag{4.2}$$

we realise that the eigenstates for the above matrix can be written in a general form of $\Psi_n = (\psi_n, a\psi_{n-2})$, with some complex coefficient a. In fact, after calculation we

---

[1] From now on, we supress the index q as irrelevant to our work.

[2] We recall here that the basis is constructed as $(\phi_{+,A1}, \phi_{+,B2})^T$ in the K_+ and $(\phi_{-,B2}, \phi_{-,A1})^T$ in the K_- valley, see Sect. 2.3.

obtain a set of eigenvalues and eigenfunctions

$$\epsilon_{n\alpha} = \alpha \frac{2\hbar^2 v^2}{\gamma_1 \lambda_B^2} \sqrt{n(n-1)}; \tag{4.3}$$

$$\Psi_{n\alpha} = \begin{pmatrix} \psi_n \\ 0 \end{pmatrix}, \; n = 0, 1; \quad \Psi_{n\alpha} = \frac{1}{\sqrt{2}} \begin{pmatrix} \psi_n \\ \alpha \psi_{n-2} \end{pmatrix}, \; n \geq 2; \tag{4.4}$$

where we use the index n to number the Landau levels and α to distinguish between the valence ($\alpha = -1$) and conduction ($\alpha = 1$) band. We see that at low energies, the energy of the Landau level is proportional to the strength of the magnetic field B. Also, for large n the LLs are almost equidistant as $\sqrt{n(n-1)} \approx n$. Two levels, $n = 0$ and $n = 1$, have the same energy, $\epsilon_0 = \epsilon_1 = 0$, giving rise to an unusual, 8-fold degenerate Landau level, which is shared between electrons and holes and thus does not require any index α. Interestingly, for $n = 0, 1$, the wavefunction posesses a component only in one of the layers: layer 1 for the electronic states in the valley K_+ and layer 2 for the states in the valley K_-.

4.1.2 Landau Levels in the Four-Band Model

The derivation of the Landau level spectra within the four-band model follows very much the same approach as in the case of the two-band approximation. Again, we neglect γ_3 and other less important couplings and we start with the neutral bilayer. In this case, we construct the eigenstates using functions ψ_n and ψ_{n-2} (for the first two components of the eigenstate), as well as ψ_{n-1} (for the last two components). In the following, we use superscript c (s) to denote Landau levels originating from the low-energy (high-energy) bands and α equal to 1 or -1 to indicate the sign of the energy. The exact energies for the low-energy bands are given by

$$\epsilon_0^c = \epsilon_1^c = 0,$$

$$\epsilon_{n\alpha}^c = \frac{\alpha}{\sqrt{2}} \left(\gamma_1^2 + 2\frac{\hbar^2 v^2}{\lambda_B^2}(2n-1) - \sqrt{\gamma_1^4 + \frac{4\hbar^2 v^2 \gamma_1^2}{\lambda_B^2}(2n-1) + \frac{4\hbar^4 v^4}{\lambda_B^4}} \right)^{\frac{1}{2}}$$
$$\text{for } n \geq 2. \tag{4.5}$$

The corresponding expression for the high-energy bands is

$$\epsilon_{n\alpha}^s = \frac{\alpha}{\sqrt{2}} \left(\gamma_1^2 + 2\frac{\hbar^2 v^2}{\lambda_B^2}(2n-1) + \sqrt{\gamma_1^4 + \frac{4\hbar^2 v^2 \gamma_1^2}{\lambda_B^2}(2n-1) + \frac{4\hbar^4 v^4}{\lambda_B^4}} \right)^{\frac{1}{2}}$$
$$\text{for } n \geq 1. \tag{4.6}$$

Note that in this formulation, for high-energy LLs indexing starts with $n = 1$, not $n = 0$, emphasizing the distinctiveness of the $\epsilon = 0$ LL. Each level has additional 4-fold degeneracy due to valleys and spins.

For magnetic fields B<20T, expressions for the LL energy in a symmetric bilayer, Eqs. (4.5) and (4.6), can be simplified using a small parameter $x = \frac{\sqrt{2}\hbar v}{\lambda_B \gamma_1}$; $|x| \ll 1$. We obtain the following approximations for the energies $\epsilon_{n\alpha}^c$ and $\epsilon_{n\alpha}^s$:

$$\epsilon_{n\alpha}^c \approx \alpha \gamma_1 \sqrt{n(n-1)} x^2 \left[1 - (n - \tfrac{1}{2})x^2 \right],$$

$$(4.7)$$

$$\epsilon_{n\alpha}^s \approx \alpha \gamma_1 \left[1 + \tfrac{1}{2}(2n-1)x^2 - (n^2 - n + \tfrac{1}{8})x^4 \right].$$

Equations (4.7) allow us to write the eigenstates in terms of the powers of x. For example, for the low-energy bands, we suggest the eigenstates in the form of:

$$\psi_{n\alpha}^c = \frac{1}{\sqrt{2}} \begin{pmatrix} \left[1 + \sum_{k=1} A_k x^k\right]\psi_n \\ \alpha\left[1 + \sum_{k=1} B_k x^k\right]\psi_{n-2} \\ i\xi\sqrt{n}x\left[1 + \sum_{k=1} C_{k+1}x^k\right]\psi_{n-1} \\ -i\xi\alpha\sqrt{n-1}x\left[1 + \sum_{k=1} D_{k+1}x^k\right]\psi_{n-1} \end{pmatrix}, \quad n \geq 2, \qquad (4.8)$$

what leads to the following relations between the coefficients[3]:

$$\begin{aligned} B_k &= A_k + \tfrac{1}{2}A_{k-2}, & B_1 &= A_1, \\ C_{k+1} &= A_k - (n-1)A_{k-2}, & C_2 &= A_1, \\ D_{k+1} &= A_k - (n-\tfrac{1}{2})A_{k-2}, & D_2 &= A_1. \end{aligned}$$

Keeping terms up to x^2 and requiring that the eigenstates are normalised, we have

$$\psi_{n\alpha}^c = \frac{1}{\sqrt{2}} \begin{pmatrix} \left[1 - \tfrac{1}{2}nx^2\right]\psi_n \\ \alpha\left[1 - \tfrac{1}{2}(n-1)x^2\right]\psi_{n-2} \\ i\xi\sqrt{n}x\psi_{n-1} \\ -i\xi\alpha\sqrt{n-1}x\psi_{n-1} \end{pmatrix}, \quad n \geq 2, \qquad (4.9)$$

and the two 'special' eigenstates ($n = 0$ and $n = 1$),

$$\psi_0 = \begin{pmatrix} \psi_0 \\ 0 \\ 0 \\ 0 \end{pmatrix}, \quad \psi_1^c = \begin{pmatrix} \left[1 - \tfrac{1}{2}x^2\right]\psi_1 \\ 0 \\ i\xi x\psi_0 \\ 0 \end{pmatrix}. \qquad (4.10)$$

For the high-energy bands, we get:

---

[3] These relations were not investigated beyond $k = 3$.

$$\psi_{n\alpha}^s \approx \frac{1}{\sqrt{2}} \begin{pmatrix} i\xi x \sqrt{n}\psi_n \\ -i\xi\alpha x\sqrt{n-1}\psi_{n-2} \\ \left(1 - \frac{nx^2}{2}\right)\psi_{n-1} \\ \alpha\left[1 - \frac{x^2}{2}(n-1)\right]\psi_{n-1} \end{pmatrix}, n \geq 1. \tag{4.11}$$

We now turn to asymmetric bilayer graphene. In particular, we are interested in (i) the interlayer asymmetry u and (ii) the substrate-induced intralayer asymmetry in the bottom layer only, δ. The former is taken exactly as defined in Sect. 2.2.3 while the latter can be obtained by taking $\Delta_{AB} = \Delta$, and reflects in the simplest way interaction of bilayer with an underlying substrate. In the case of bilayer graphene grown epitaxially on diatomic substrate, such as SiC, this interaction distinguishes between sublattices in the bottom layer. However, due to a much greater distance, the interaction between the substrate and the top layer is neglected. The Hamiltonian is

$$\hat{H} = \xi \begin{pmatrix} \frac{u}{2} + \frac{\delta}{4}(1+\xi) & 0 & 0 & v\hat{\pi}^\dagger \\ 0 & -\frac{u}{2} - \frac{\delta}{4}(1-\xi) & v\hat{\pi} & 0 \\ 0 & v\hat{\pi}^\dagger & -\frac{u}{2} + \frac{\delta}{4}(1-\xi) & \xi\gamma_1 \\ v\hat{\pi} & 0 & \xi\gamma_1 & \frac{u}{2} - \frac{\delta}{4}(1+\xi) \end{pmatrix}. \tag{4.12}$$

Let us treat the diagonal part as a small perturbation and concentrate on Landau levels $n = 0, 1$. Then, using standard perturbation theory, we can obtain the following changes of the energies ϵ_0 and ϵ_1^c,

$$\epsilon_0 = \xi\frac{u}{2} + \xi\frac{\delta}{4}(1+\xi),$$
$$\epsilon_1^c = \epsilon_0 - x^2\left(\xi u + \frac{1}{2}\delta\right). \tag{4.13}$$

Interestingly, the energy ϵ_1^c for one of the valleys decreases with magnetic field, whereas ϵ_0 is independent of B. As a result, Landau level with index $n = 1$ is closer to the neutrality point than the Landau level $n = 0$.

To find the perturbed eigenstates, we notice that because the perturbation is diagonal, only mixing between Landau levels with the same Landau index is important. As a result, the ψ_0^c state remains unperturbed, while

$$\psi_1^c \approx \begin{pmatrix} \left[1 - \frac{1}{2}x^2\right]\psi_1 \\ 0 \\ i\xi x\psi_0 \\ i\xi\frac{x}{2\gamma_1}(2\xi u + \delta)\psi_0 \end{pmatrix},$$

$$\psi_{1\alpha}^s \approx \frac{1}{\sqrt{2}} \begin{pmatrix} i\xi x \left(1 + \frac{1}{\gamma_1}\left[\xi u \left(1 - \frac{1}{4}\alpha\right) + \frac{\delta}{2}\left(1 + \frac{1}{4}\alpha\xi\right)\right]\right)\psi_1 \\ 0 \\ \left[1 - \frac{x^2}{2} - \frac{\alpha}{2\gamma_1}\left(\xi\frac{u}{2} - \xi\frac{\delta}{4}\right)\right]\psi_0 \\ \alpha\left[1 + \frac{\alpha}{2\gamma_1}\left(\xi\frac{u}{2} - \xi\frac{\delta}{4}\right)\right]\psi_0 \end{pmatrix},$$

where we kept only the leading terms in the perturbation and x.

Similar calculation can be performed for the high-energy states for $n \geq 2$, yielding energies

$$\epsilon_n^s \approx \alpha\gamma_1\left[1 + \frac{1}{2}(2n - 1)x^2 - \left(n^2 - n + \frac{1}{8}\right)x^4\right] - \frac{\delta}{4}$$
$$+ x^2\left[\frac{\xi u}{2} + \frac{\delta}{2}\left(n - \frac{1}{2}\right)\right], \quad n \geq 1,$$

and perturbed eigenstates

$$\psi_{n\alpha}^s \approx \frac{1}{\sqrt{2}} \begin{pmatrix} i\xi x\sqrt{n}\left[1 + \frac{\xi\alpha}{\gamma_1}\left(\frac{7}{4}u + \delta\left(\xi + \frac{1}{8}\right)\right)\right]\psi_n \\ -i\xi\alpha x\sqrt{n-1}\left[1 - \frac{\xi\alpha}{\gamma_1}\left(\frac{7}{4}u - \delta\left(\xi - \frac{1}{8}\right)\right)\right]\psi_{n-2} \\ \left[1 - \frac{1}{2}nx^2 - \frac{\xi\alpha}{4\gamma_1}\left(u - \frac{\delta}{2}\right)\right]\psi_{n-1} \\ \alpha\left[1 - \frac{1}{2}(n-1)x^2 + \frac{\xi\alpha}{4\gamma_1}\left(u - \frac{\delta}{2}\right)\right]\psi_{n-1} \end{pmatrix}, \quad n \geq 2. \tag{4.14}$$

However, the change of the low-energy Landau levels with $n \geq 2$ is more difficult to describe. Because of the closeness of two of the levels, perturbation theory is only applicable if u and δ are small in comparison to the spacing between them, $u, \delta \ll 2\gamma_1 x^2\sqrt{n(n-1)}$ (this translates into the scaling $\frac{u}{B}, \frac{\delta}{B} \ll 4.6\frac{\text{meV}}{\text{T}}$). Alternatively, if $u, \delta \sim \gamma_1\sqrt{n(n-1)}x^2 \ll \gamma_1$, one can use the effective Hamiltonian,

$$\hat{H}_{\text{eff}} = -\frac{v^2}{\gamma_1}\begin{pmatrix} 0 & \left(\hat{\pi}^\dagger\right)^2 \\ \hat{\pi}^2 & 0 \end{pmatrix} + \begin{pmatrix} \xi\frac{u}{2} + \xi\frac{\delta}{4}(1+\xi) & 0 \\ 0 & -\xi\frac{u}{2} - \xi\frac{\delta}{4}(1 - xi) \end{pmatrix}$$
$$+ \frac{v^2}{\gamma_1^2}\begin{pmatrix} -\left(\xi u + \frac{\delta}{2}\right)\hat{\pi}^\dagger\hat{\pi} & 0 \\ 0 & \left(\xi u - \frac{\delta}{2}\right)\hat{\pi}\hat{\pi}^\dagger \end{pmatrix}. \tag{4.15}$$

This Hamiltonian has been obtained in a manner described in Sect. 2.3. The resulting energies and eigenstates are

$$\epsilon_{n\alpha}^c \approx \frac{\delta}{4} - x^2\left[\xi\frac{u}{2} + \frac{\delta}{2}\left(n - \frac{1}{2}\right)\right] + \alpha\sqrt{(\epsilon_n)^2 + \frac{1}{4}\left(u^2 + u\delta + \frac{1}{16}\delta^2\right)},$$

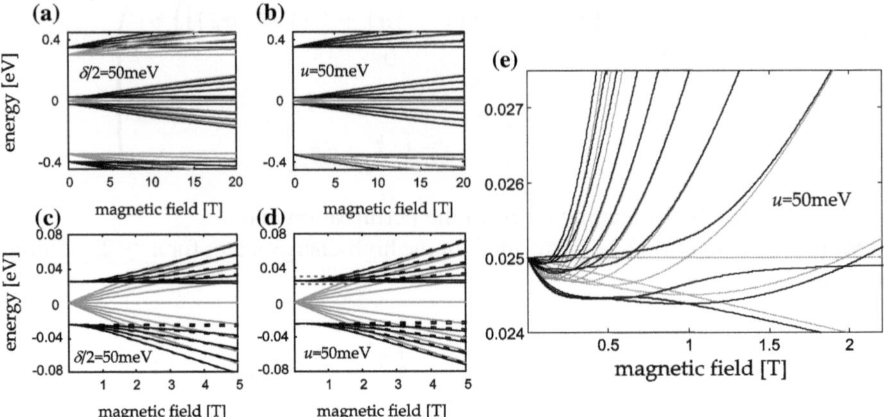

Fig. 4.1 Numerically calculated Landau levels of bilayer graphene for range of high (**a** and **b**) and low (**c** and **d**) energies; *black solid* and *dashed lines* in **c** and **d** represent levels at valley K_+ and K_- respectively, grey solid lines in the background show the Landau level spectrum for no asymmetries ($u, \delta = 0$); in **a** and **b** only the spectrum at K_+ is shown for clarity. Note that nonzero u in (**b**) affects the high-energy LLs very weakly and so the corresponding grey lines are underneath the black ones. Zero of the energy scale is shifted to the middle of the gap opened in each case at the K point. **e** Area bounded by dashed red rectangle in **d** shown again with the 11 lowest LLs at K+: black solid lines - taking into account $v_3 = 0.15 eV$, grey dashed lines - v_3 neglected. Values of parameters used: $v = 1 \times 10^6$ m/s, $\gamma_1 = 0.35$ eV, resulting in low-energy effective mass $m_{\text{eff}} = \gamma_1/2v^2 \approx 0.03$ of the electron mass. Figure reprinted from Ref. [28], Copyright (2009), with permission from IOP Publishing

$$\tilde{\psi}_{n\alpha}^c = \frac{1}{\sqrt{C}} \left(\begin{matrix} \epsilon_n \psi_n \\ \left(\tilde{\epsilon}_{n\alpha}^c - \xi \left[\frac{u}{2} + \frac{\delta}{4}(1 + \xi)\right] + \left(\xi u + \frac{\delta}{2}\right) n x^2\right) \psi_{n-2} \end{matrix} \right).$$

Above, $\epsilon_n = \gamma_1 x^2 \sqrt{n(n-1)}$, C is the normalisation coefficient and we used the tilde symbol to temporarily distinguish the eigenstate of the effective two-band Hamiltonian from the eigenstate of the four-band model. We can recover the high-energy components of the eigenstate, to obtain

$$\psi_{n\alpha}^c \approx \frac{1}{\sqrt{C}} \left(\begin{matrix} \epsilon_n \psi_n \\ \left(\epsilon_{n\alpha}^c - \xi \left[\frac{u}{2} + \frac{\delta}{4}(1 + \xi)\right] + \left(\xi u + \frac{\delta}{2}\right) n x^2\right) \psi_{n-2} \\ i\xi x \epsilon_n \sqrt{n} \psi_{n-1} \\ -i\xi x \sqrt{n-1} \left(\epsilon_{n\alpha}^c - \xi \left[\frac{u}{2} + \frac{\delta}{4}(1 + \xi)\right] + \left(\xi u + \frac{\delta}{2}\right) n x^2\right) \psi_{n-1} \end{matrix} \right).$$

$$\tag{4.16}$$

The numerically calculated Landau level spectra both in the absence and presence of the asymmetries u and $\delta/2$ are shown in Fig. 4.1a–d. The low-energy Landau level spectrum for neutral bilayer, equivalent with Eq. (4.5), is shown as grey solid lines in Fig. 4.1a–d. Those levels create a fan-plot originating at zero energy. As shown earlier within the two-band approximation, levels $n = 0$ and $n = 1$ have the same energy, leading to an unusual 8-fold degenerate level at $\epsilon = 0$. The high-energy LLs, Eq. (4.6), create two additional fan-plots originating at $\epsilon = \pm\gamma_1$, which are shown together with low-energy LLs with grey solid lines in Fig. 4.1a–b. The Landau level spectra for $u = 50\,\mathrm{meV}$ ($\frac{\delta}{2} = 50\,\mathrm{meV}$) is shown with black lines in Fig. 4.1b and d (a and c). Both of the asymmetries lift the valley degeneracy. Also, in both cases the additional degeneracy of $n = 0$ and $n = 1$ LLs is removed, $\epsilon_1 \neq \epsilon_0$ [note that the $n = 0$ level is shifted with respect to the middle of the gap at the K point from $\epsilon = 0$ in opposite directions in the valley K_+ ($\xi = +$) and K_- ($\xi = -$)]. As can be seen in Fig. 4.1a, interlayer asymmetry u affects the high-energy LLs very weakly and so, the corresponding black lines cover grey lines showing high-energy LLs in the symmetric bilayer. At low energies and low fields (Fig. 4.1e), signatures of a Mexican hat developing in the electronic spectrum of an asymmetric bilayer can be noticed in the fan-plots of the LL spectrum. Inverted curvature in the central part of such a structure (hole-like in conduction and electron-like in valence band) results in inverted behaviour of Landau levels at very low B [the energy of electron (hole) levels decrease (increase) with increasing B] which then returns to typical behaviour at higher B [the energy of electron (hole) levels increase (decrease) with increasing B]. This results in interlevel crossings. Also, this is a regime where the influence of the parameter v3, neglected so far, is important, because it mixes LLs n and $n - 3$, thus changing some of the interlevel crossings into anticrossings. An example of the numerically calculated spectrum taking into account v_3 using a procedure explained in the next section is shown in Fig. 4.1e.

4.1.3 Numerical Treatment of the γ_3 Coupling

The derivation of the Landau level structure of bilayer graphene around a single valley, as shown in previous sections, is possible because each of the components of the wave function can be written in terms of only one Landau function ψ_n. Any of the on-site asymmetries, as well as the couplings γ_0, γ_1 or γ_4 can be incorporated into this scheme. However, the γ_3 coupling leads to mixing of the LLs with each other and the problem can no longer be solved in this way. This perturbation becomes important only at weak magnetic fields and in most situations, it can be neglected for fields $B > 5\mathrm{T}$. Nevertheless, we investigate here how the γ_3 terms affect the Landau level structure of bilayer graphene at weak magnetic fields. This is because with improving sample fabrication processes, it is now possible to probe the physics at energies where γ_3 is important [30]. We use an approach similar to that developed to treat γ_3 in the calculation of the LL spectrum of graphite [25] and for simplicity, discuss the case of the two-band model, Eq. (2.12).

In an external magnetic field, the γ_3 perturbation mixes LLs with each other so that the electron amplitude on each sublattices is now a linear combination of infinitely many functions ψ_n. Using the knowledge from the previous sections, we again want to describe the eigenstates with vectors whose odd and even components correspond to electron wave amplitudes on one of the sublattices, $B2$ ($A1$) and $A1$ ($B2$) in the K_+ (K_-) valley, respectively. Also, we want each component to be expressed using

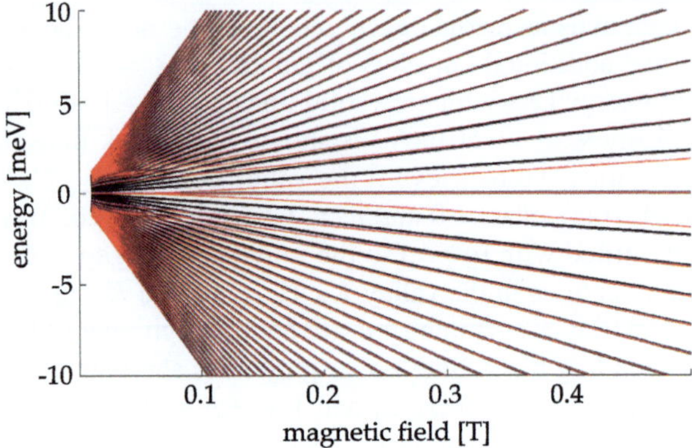

Fig. 4.2 Comparison of the low-energy and weak-field Landau level structures obtained neglecting the γ_3 coupling (*black lines*) and taking it into account (*red lines*). Values of parameters used: $v = 1 \times 10^6$ m/s, $\gamma_1 = 0.4$ eV, resulting in low-energy effective mass $m_{\text{eff}} = \gamma_1/2v^2 \approx 0.03$ of the electron mass, $v_3 = 0.1v$ (if applicable)

a single function ψ_n. We choose the ordering of the entries corresponding to specific ψ_n in such a way, so that pair of vectors $(2n - 1)$-th and $(2n)$-th create a minimal subspace required to describe the n-th LL in the absence of γ_3. In the absence of γ_3, this tallies to repeating Hamiltonian (4.2) as the diagonal block of an infinite matrix, with all other elements equal to zero. The presence of γ_3 leads to some off-diagonal perturbations in this matrix, which can be written in the form[4]:

$$\hat{H} = \begin{pmatrix} 0 & 0 & 0 & \hat{D}(1) & 0 & \cdots \\ 0 & 0 & 0 & 0 & \hat{D}(2) & \cdots \\ 0 & 0 & \hat{H}(1) & 0 & 0 & \cdots \\ \hat{D}^\dagger(1) & 0 & 0 & \hat{H}(2) & 0 & \cdots \\ 0 & \hat{D}^\dagger(2) & 0 & 0 & \hat{H}(3) & \cdots \\ \vdots & \vdots & \vdots & \vdots & \vdots & \ddots \end{pmatrix}, \tag{4.17}$$

where

$$\hat{H}(n) = \begin{pmatrix} 0 & \frac{v^2}{\gamma_1}x^2\sqrt{n(n+1)} \\ \frac{v^2}{\gamma_1}x^2\sqrt{n(n+1)} & 0 \end{pmatrix},$$

$$\hat{D}(n) = \begin{pmatrix} 0 & -i\xi x v_3 \sqrt{n} \\ 0 & 0 \end{pmatrix}. \tag{4.18}$$

---

[4] Note, that two rows and columns in the following matrix, identically equal to zero, give rise to two solutions at zero energy, which correspond to unphysical eigenstates (zero vectors) and should not be confused with the zero energy Landau levels described earlier in the text.

We then truncate the infinite basis, restricting the calculation to a given n LLs and diagonalise the resulting Hamiltonian numerically. The number of the basis vectors required in the calculation in order to describe properly the low-energy LL structure increases with decreasing magnetic field (reflecting growing importance of the γ_3 terms at weaker fields). Basis of of the dimension 800 is enough to describe LL spectra for magnetic fields $B > 0.01\mathrm{T}$. Similar analysis can be performed within the four-band model. However, the matrix dimension has to be then doubled for the same accuracy at low energies, whereas the correction to the high energy LLs is negligible.

The low-energy Landau level spectra at weak fields has been presented in Fig. 4.2. The *black lines* show the LL spectrum equivalent to that described in Eq. (4.3), whereas the red lines demonstrate the low-energy LL spectrum with the γ_3 taken into account (via a procedure explained above). Only the energies of few lowest Landau levels are affected. The unusual degeneracy of the $\epsilon = 0$ Landau level is preserved down to the lowest fields, where two LLs merge with the zero-energy level. Then, the $n = 0$ LL becomes 16-fold degenerate reflecting the existence of four Dirac cones in the electronic spectrum.

4.2 Magneto-Optical Selection Rules and the Absorption Spectra

We describe the electron-photon interaction within the four-band linear approximation, by expanding the Hamiltonian $\hat{H}_\xi(\boldsymbol{p} - e\boldsymbol{A})$, where $[\boldsymbol{p} - e\boldsymbol{A}]$ is the canonical momentum including the electromagnetic vector potential $\boldsymbol{A}$, up to the first power in $\boldsymbol{A}$, and write the interaction Hamiltonian

$$\hat{H}_{\mathrm{int}} = \hat{\boldsymbol{j}} \cdot \boldsymbol{A}, \tag{4.19}$$

where $\hat{\boldsymbol{j}} = -e(\frac{\partial \hat{H}_\xi}{\partial p_x}, \frac{\partial \hat{H}_\xi}{\partial p_y})$ is the current operator. The incoming beam is characterised by a time-dependent electric field, $\boldsymbol{E}_\omega(t) = \boldsymbol{E}_\omega e^{-i\omega t}$. Using Maxwell's equations, we arrive at

$$\boldsymbol{A} = \frac{1}{-i\omega}\boldsymbol{E}_\omega e^{-i\omega t}. \tag{4.20}$$

Wave functions derived in Sect. 4.1.2 can now be used to determine transition rules for the absorption of right ($\oplus$) and left-handed ($\ominus$) circularly polarized light $\boldsymbol{E}_\omega = E_\omega l_{\oplus/\ominus}$, with $l_\oplus = \frac{1}{\sqrt{2}}(l_x - il_y)$ and $l_\ominus = \frac{1}{\sqrt{2}}(l_x + il_y)$. Neglecting for now the prefactor $\frac{E_\omega}{-i\omega}$ in the interaction above, we end with perturbation $\left(\hat{\boldsymbol{j}} \cdot l_{\oplus/\ominus}\right)$ for circularly polarised light interacting with electrons in the material. We can now find optical strengths of inter-LL transitions. For the transitions between the low-energy bands ($c \to c$), we obtain:

$$|\langle\psi_1^c|\hat{\boldsymbol{j}}\cdot\boldsymbol{l}_\oplus|\psi_0^c\rangle\|^2 = |\langle\psi_0^c|\hat{\boldsymbol{j}}\cdot\boldsymbol{l}_\ominus|\psi_1^c\rangle|^2 \approx \frac{e^2v^2x^2(2\xi u+\delta)^2}{2\gamma_1^2},$$

$$|\langle\psi_{2\alpha}^c|\hat{\boldsymbol{j}}\cdot\boldsymbol{l}_\oplus|\psi_1^c\rangle\|^2 = |\langle\psi_1^c|\hat{\boldsymbol{j}}\cdot\boldsymbol{l}_\ominus|\psi_{2\alpha}^c\rangle\|^2$$

$$\approx \frac{8e^2v^2x^2}{C}\left\{\epsilon_{2\alpha}^c - \xi\left[\frac{u}{2}+\frac{\delta}{4}(1+\xi)\right]+(2\xi u+\delta)\,x^2\right\}^2, \qquad (4.21\mathrm{a})$$

$$|\langle\psi_{n\beta}^c|\hat{\boldsymbol{j}}\cdot\boldsymbol{l}_\oplus|\psi_{m\alpha}^c\rangle\|^2 = |\langle\psi_{m\alpha}^c|\hat{\boldsymbol{j}}\cdot\boldsymbol{l}_\ominus|\psi_{n\beta}^c\rangle\|^2$$

$$\approx \delta_{n-1,m}\frac{2me^2v^2x^2}{C^2}\epsilon_m^2\left\{\epsilon_{n\beta}^c - \xi\left[\frac{u}{2}+\frac{\delta}{4}(1+\xi)\right]+\left(\xi u+\frac{\delta}{2}\right)nx^2\right\}^2.$$

For the transitions between the low-energy and the high-energy bands ($s \leftrightarrow c$), we have:

$$|\langle\psi_{n\alpha}^s|\hat{\boldsymbol{j}}\cdot\boldsymbol{l}_\oplus|\psi_m^c\rangle|^2 = |\langle\psi_m^c|\hat{\boldsymbol{j}}\cdot\boldsymbol{l}_\ominus|\psi_{n\alpha}^s\rangle|^2 \approx e^2v^2\delta_{n-1,m}, m = 0, 1,$$

$$|\langle\psi_{n\alpha}^s|\hat{\boldsymbol{j}}\cdot\boldsymbol{l}_\oplus|\psi_{m\beta}^c\rangle|^2 = |\langle\psi_{m\beta}^c|\hat{\boldsymbol{j}}\cdot\boldsymbol{l}_\ominus|\psi_{n\alpha}^s\rangle|^2$$

$$\approx \frac{e^2v^2\epsilon_m^2}{C}\left[1+\frac{\xi\alpha}{4\gamma_1}(2u-\delta)-3mx^2\right]\delta_{n-1,m}, m \geq 2, \qquad (4.21\mathrm{b})$$

$$|\langle\psi_{n\beta}^c|\hat{\boldsymbol{j}}\cdot\boldsymbol{l}_\oplus|\psi_{m\alpha}^s\rangle|^2 = |\langle\psi_{m\alpha}^s|\hat{\boldsymbol{j}}\cdot\boldsymbol{l}_\ominus|\psi_{n\beta}^c\rangle|^2$$

$$\approx \frac{e^2v^2(C-\epsilon_n^2)}{C}\left[1-\frac{\xi\alpha}{4\gamma_1}(2u-\delta)-3mx^2\right]\delta_{n-1,m}, m \geq 1.$$

Finally, for the transitions among the high-energy bands alone ($s \rightarrow s$), we have:

$$|\langle\psi_{n\beta}^s|\hat{\boldsymbol{j}}\cdot\boldsymbol{l}_\oplus|\psi_{m\alpha}^s\rangle|^2 = |\langle\psi_{m\alpha}^s|\hat{\boldsymbol{j}}\cdot\boldsymbol{l}_\ominus|\psi_{n\beta}^s\rangle|^2 \approx 2e^2v^2x^2m\delta_{n-1,m}, m \geq 1. \quad (4.21\mathrm{c})$$

Equations in 4.21 generalise the earlier study of optical and magneto-optical absorption in bilayers [22]. Examples of allowed transitions are illustrated in Fig. 4.3a, b. Independently of the presence/absence of asymmetries δ or u, selection rules for absorption of right-handed polarized light are such that the Landau level index has to be decreased by one, whereas absorption of left-handed photons requires an increase of the Landau level index by one. Also, optical strengths of $c \rightarrow c$ and $s \rightarrow s$ transitions are proportional to the magnetic field B and LL index, whereas optical strengths of $s \leftrightarrow c$ transitions are almost independent of B. As x is a small parameter, the intensity of the first $s \leftrightarrow c$ transitions should be higher than for $c \rightarrow c$ transitions corresponding to the same energy of incident radiation $\hbar\omega$.

We describe the optical absorption of the incoming photons by the sample in the presence of an external magnetic field, I_{abs}, by the ratio of the energy absorbed by the material to the energy carried by the electromagnetic field, $\langle S\rangle \propto E_\omega^2$. The energy absorbed is proportional to the energy of the incoming photons, density of electrons in the Landau level (and thus to the strength of the magnetic field B) and the transition probability between the initial and final states, expressed by the optical

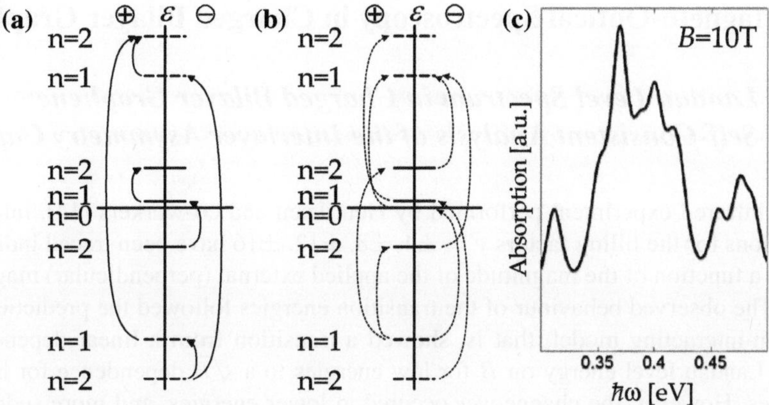

Fig. 4.3 Allowed intraband (**a**) and interband (**b**) optical excitations. Low-energy (high-energy) LLs are depicted with *solid (dashed) lines*. Transitions in $\mathbf{l}_\oplus$ ($\mathbf{l}_\ominus$) are shown on the *left (right)* of the energy axis, under $\oplus$ ($\ominus$) symbol. Energy axis not to scale; not shown is very weak $|0_c\rangle \rightarrow |1_c\rangle$ transition allowed only in the presence of asymmetry. **c** Magneto-optical absorption spectrum for the energy of incident light $\hbar\omega \approx |\gamma_1|$; magnetic field $B = 10$T, $\gamma_1 = -0.35$ eV, $v = 10^6$ m/s; Landau level broadening was approximated with a Lorentzian with full width at half maximum 20 meV. Figure reprinted from Ref. [28], Copyright (2009), with permission from IOP Publishing

strengths above. One also has to take into account electron level occupancy f_n of the level $|n\rangle$ and conservation of energy in the process. Hence, the intensity of absorption, I_{abs}, is proportional to

$$I_{\text{abs}}(\omega) \propto \sum_{m,n} \frac{(\epsilon_n - \epsilon_m)\,(f_m - f_n)\,\left|\langle n|\hat{\boldsymbol{j}} \cdot \boldsymbol{l}_{\oplus/\ominus}|m\rangle\right|^2\,B}{\omega^2}\delta(\epsilon_n - \epsilon_m - \hbar\omega), \quad (4.22)$$

where $|m\rangle$ and $|n\rangle$ are the initial and final electron states, respectively. In Fig. 4.3c, the numerically calculated magneto-optical spectrum of symmetric bilayer for $\hbar\omega \approx |\gamma_1|$ has been shown (the range of energies at which the $s \leftrightarrow c$ transitions become possible). For this purpose, we approximated the Dirac delta expressing the energy conservation with a Lorentzian, with the full width at half-maximum parameter Γ associated with the broadening of the Landau levels (as the incoming photon beam is monochromatic). We assumed the same broadening of all Landau levels. The onset of $s \leftrightarrow c$ transitions (the two highest peaks around $\hbar\omega \approx 0.39$ eV) can be observed against the background of $c \rightarrow c$ excitations.

4.3 Magneto-Optical Spectroscopy in Charged Bilayer Graphene

4.3.1 Landau Level Spectrum in Charged Bilayer Graphene: Self-Consistent Analysis of the Interlayer Asymmetry Gap

In the infrared experiment performed by Henriksen and co-workers [14], inter-LL transitions for the filling factors $\nu = \pm 4, \pm 8, \pm 12, \pm 16$ have been traced individually as a function of the magnitude of the applied external (perpendicular) magnetic field. The observed behaviour of the transition energies followed the predictions of the non-interacting model, that is, showed a transition from a linear dependence of the Landau level energy on B for low energies to a $\sqrt{B}$ dependence for higher energies. However, the changeover occured at lower energies, and more suddenly, than expected. For some filling factors, a better fit was actually achieved when the data was fitted to monolayer dispersion rather than bilayer one. In their conclusion, the authors suggested many-body interactions as responsible for this departure from the predictions of the single-particle theory. However, the experimental setup and the need to fix the filling factor to trace a particular transition while the magnetic field was changed, implies significant changes of the carrier density in bilayer graphene during the experiment. That was achieved by applying an external electric field perpendicular to the layers. Such external electric field is known to induce interlayer asymmetry in the system [31–33], in our model described by the on-site energy u (see Sect. 2.2.3). As has been shown in the previous sections, non-zero asymmetry u, caused by a possible difference in electric potential energy between the layers, opens a gap in the electronic spectrum and, in the presence of an additional external magnetic field, modifies the LL spectrum [26, 28, 34–36], thus affecting inter-LL transition energies. To model this effect, we employ a self-consistent theory of the charging of bilayer graphene in external magnetic field. We extend here the self-consistent analysis of Ref. [31], performed in the absence of the magnetic field, into the regime of strong quantizing magnetic fields, taking into account the possibility that there is a finite asymmetry already in a neutral structure (see Eq. 4.25 below). Our investigation has first been published in Ref. [29].

Let us consider a gated bilayer as shown in Fig. 4.4. The interlayer distance is c_0. In an external magnetic field B, the Landau levels are described by the four-band Hamiltonian, Eq. (4.12). We neglect from now on the intralayer asymmetry δ. In order to keep the filling factor ν fixed while changing B, a total excess density, $n = \nu \frac{eB}{h}$, must be induced using the gate. The density n is shared between the two layers: $n = n_1 + n_2$, where, assuming a top gate, n_2 (n_1) is the excess density on the layer closest to (furthest from) the gate. The interlayer asymmetry u is a result of different electric potentials U_1^{pot} and U_2^{pot} on the first and second layer, respectively,

$$u = U_1^{\mathrm{pot}} - U_2^{\mathrm{pot}}. \tag{4.23}$$

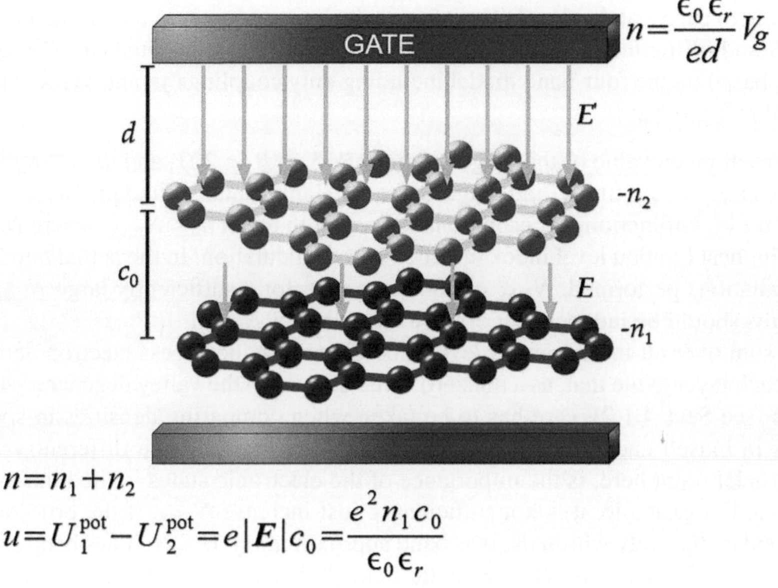

$$n = n_1 + n_2$$
$$u = U_1^{\mathrm{pot}} - U_2^{\mathrm{pot}} = e|E|c_0 = \frac{e^2 n_1 c_0}{\epsilon_0 \epsilon_r}$$

Fig. 4.4 Schematic of a 'biased bilayer' showing all charge densities and electric fields induced by a single gate

We assume that there is no electric field below the bottom layer (layer 1). Then, by applying Gauss' law, we can notice that the electric field E between the layers, which arises due to the incomplete screening of the gate electric field by the charge en_2 on the top layer, is related to the unscreened charge density n_1,

$$|\boldsymbol{E}| = \frac{en_1}{\epsilon_0 \epsilon_r}, \tag{4.24}$$

where here ϵ_0 and ϵ_r are the permittivity of vacuum and the effective dielectric constant determined by the substrate (in the actual experiment, bilayer graphene was deposited on SiO_2), respectively. Finally, using the relation between the electric field and electric potential, we can write the equation relating the interlayer asymmetry u to the density n_1,

$$u(\nu, B) = w + \frac{e^2 c_0 n_1(\nu, B)}{\epsilon_0 \epsilon_r}. \tag{4.25}$$

Here, w takes into account a finite asymmetry of a neutral structure (internal electric field due to, for example, initial non-intentional doping of the flake by deposits/adsorbates). In our numerical calculations we use $\epsilon_r = 2$.

On the one hand, u influences the LL spectrum via the Hamiltonian in Eq. (4.12). On the other hand, its value depends on the charge density n_1, which can only be obtained with a full knowledge of the LL spectrum and the wave functions corresponding to each Landau level. Therefore, calculation of u given a specific magnetic field B and filling factor v requires a self-consistent numerical analysis. Our calculation, based on the four-band model including only couplings γ_0 and γ_1, consists of the following steps:

- For each given value of the magnetic field B, $5 < B < 20$T, and the filling factor v we choose a starting u, and diagonalize the Hamiltonian to find the LL spectrum and the eigenfunctions for each Landau level with index $n \leq N_{\mathrm{max}}$, where N_{max} is the highest Landau level index included in the calculation. In the actual numerical calculations performed, $N_{\mathrm{max}} \sim 300$ (note that, for a sufficiently large N_{max}, the results should be independent of the exact value of N_{max}).
- We sum over all filled Landau levels and determine the excess electron densities on each layer. Note that, as a nonzero value of u splits the valley degeneracy of the LLs (see Sect. 4.1.2), care has to be taken when comparing densities in specific LLs in biased and neutral structures, not to confuse levels in different valleys. A crucial point here, is the importance of the electronic states in the high-energy bands. For example, it is not sufficient to just increase N_{max}, if the procedure is treated exclusively within the two-band approximation. In fact, it has been already shown for the case of zero magnetic field [31], that the high-energy bands are necessary for the iterative procedure to converge. This is because introduction of the interlayer asymmetry u affects the low-energy bands in a much more drastic way than the split bands (see Fig. 2.4). Effectively, the charge distribution between the layers calculated within the two-band model is much more asymmetric than it should be and as a result the value of u may not converge. In this sense, the electron density contained in the high-energy bands stabilises the procedure.
- Finally, using Eq. (4.25) we find the asymmetry parameter and, then, iterate the numerical procedure to obtain the self-consistent value of u.

In the above procedure, we implicitly assumed that Landau levels are infinitely sharp as a function of energy (no Landau level broadening considered). We also limit this procedure to integer filling factors. In fact, only those filling factors investigated in the experiment were looked at here.

The self-consistently calculated values of u obtained for several values of the filling factor v are shown in Fig. 4.5a and b for the case of the intrinsic asymmetry $w = 0$ and $w = -100$ meV, respectively. In the case when $w = 0$, the induced interlayer asymmetry is antisymmetric with respect to the change of the filling factor from positive to negative. This is because changing the filling factor from $+v$ to $-v$ corresponds to reversing the applied electric field and inducing excess densities $-n, -n_1$ and $-n_2$ and thus reversing the sign of u. Also, with decreasing B all curves tend towards $u = 0$ ($w = 0$) and $u \approx -60$ meV ($w = -100$ meV). These values are the results of the self-consistent calculation with corresponding values of w in the absence of a magnetic field [31] and provide an independent validation of our results.

Knowing the self-consistent values of u for a specific filling factor and a range of magnetic fields, we can plot the resulting low-energy Landau level spectrum. Note that although naturally similar to the spectra discussed before, these are not the

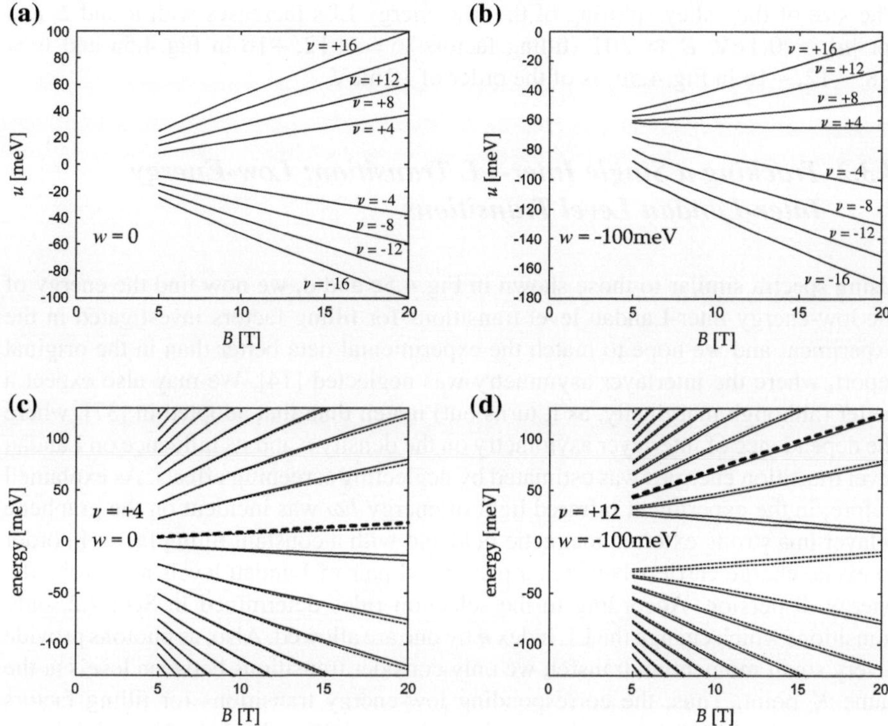

Fig. 4.5 *Top row* results of a self-consistent calculation of the interlayer asymmetry u for **a** $w = 0$ and **b** $w = -100\,$meV. *Bottom row*: the LL spectrum as a function of applied magnetic field B for constant filling factor and excess density-induced interlayer asymmetry u: **c** $\nu = +4$; $w = 0$, and **d** $\nu = +12$; $w = -100\,$meV. *Dashed* and *solid lines* denote levels belonging to K_+ and K_-, respectively. The line corresponding to the last filled Landau level is shown in *bold*. In these calculations we used $v = 10^6\,$m/s and $\gamma_1 = 0.4\,$eV. Figure reprinted from Ref. [29], Copyright (2009), with permission from Elsevier

same, as the interlayer asymmetry u takes new value for each value of B. Examples of such spectra are given in Fig. 4.5c and d for $\nu = +4$, $w = 0$ and for $\nu = +12$, $w = -100\,$meV, respectively. To refer to Landau levels as shown in Fig. 4.5, we explicitly use in what follows three symbols introduced separately before: $sn\xi$, where s attributes the LL to the conduction (+) or valence (-) band, n is the LL index and $\xi \in (+, -)$ identifies the valley (K_+ or K_-) that the level belongs to, respectively. As mentioned in the previous sections, the Landau levels $n = 0, 1$ have no s index, as those levels are shared between electrons and holes when $u = 0$. The sign of the valley splitting of the level sn depends on the sign of u: for $u > 0$, level $sn-$ has higher energy than level $sn+$ whereas the opposite is true for $u < 0$. Levels $n = 0, 1$ behave differently (see Sect. 4.1.2)—in this case

$$\left.\begin{array}{l}\epsilon_{n+} > 0 \\ \epsilon_{n-} < 0\end{array}\right\} \text{ for } u > 0, \quad \left.\begin{array}{l}\epsilon_{n+} < 0 \\ \epsilon_{n-} > 0\end{array}\right\} \text{ for } u < 0. \tag{4.26}$$

The size of the valley splitting of the low-energy LLs increases with u and B and for $|u| \approx 0.1\,\text{eV}$, $B \approx 20\text{T}$ (filling factors $v = +12, +16$ in Fig. 4.5a and $v = -8, -12, -16$ in Fig. 4.5b) is of the order of $10\,\text{meV}$.

4.3.2   Tracking a Single Inter-LL Transition: Low-Energy Inter-Landau Level Transitions

Using spectra similar to those shown in Fig. 4.5c and d, we now find the energy of the low-energy inter-Landau level transitions for filling factors investigated in the experiment and we hope to match the experimental data better than in the original report, where the interlayer asymmetry was neglected [14]. We may also expect a better (although marginally, as it turns out) match than that adopted in [37], where the dependence of interlayer asymmetry on the density n and its influence on Landau level transition energies was estimated by neglecting screening effects. As explained before, in the experiment infrared light of energy $\hbar\omega$ was incident on the graphene bilayer in a strong external magnetic field and with a constant filling factor in order to excite charge carriers between a prescribed pair of Landau levels and probe the energy dispersion. According to the selection rules determined in Sect. 4.2, only transitions which change the LL index n by one are allowed. Also, as photons provide a very small momentum transfer, we only consider transitions between levels at the same K point. Thus, the corresponding low-energy transitions for filling factors $v = +4, +8, +12, +16$ are $1\xi \to +2\xi$, $+2\xi \to +3\xi$, $+3\xi \to +4\xi$, and $+4\xi \to +5\xi$, respectively. For filling factors $v = -4, -8, -12, -16$, they are $-2\xi \to 1\xi$, $-3\xi \to -2\xi$, $-4\xi \to -3\xi$, and $-5\xi \to -4\xi$, respectively. However, as transitions between the same levels at different K points differ too little in energy to have been resolved separately in the abovementioned experiment (in fact, they can only be clearly distinguished in Fig. 4.6 for the case $v = 4$), we obtain a single transition energy $\epsilon_{\text{trans}}^{v}$ for a given filling factor v by comparing the relative intensities of the corresponding transition at each K point:

$$\epsilon_{\text{trans}}^{v} = \frac{\epsilon_{\text{trans}}^{v}(K_+)I^{v}(K_+) + \epsilon_{\text{trans}}^{v}(K_-)I^{v}(K_-)}{I^{v}(K_+) + I^{v}(K_-)} \tag{4.27}$$

where $\epsilon_{\text{trans}}^{v}(K_\xi)$ and $I^{v}(K_\xi)$ are the transition energy at the K_ξ point and its intensity, respectively. The intensities $I^{v}(K_\xi)$ are calculated as previously explained in Sect. 4.2.

The results for low-energy inter-Landau-level excitations calculated using the interlayer asymmetry $u(v, B)$ obtained in the self-consistent fashion as described in Sect. 4.3.1, are shown in Fig. 4.6 (for $w = 0$) and Fig. 4.7 (for $w = -100\,\text{meV}$). We first discuss the case $w = 0$ presented in Fig. 4.6. In this case, for a specified value of B, the asymmetry u changes sign if the sign of the filling factor is changed to the opposite (Fig. 4.5a). Hence, the Landau level spectrum for v and $-v$ are the same but the K points have to be exchanged. Therefore, $\epsilon_{\text{trans}}^{v}(K_+) = \epsilon_{\text{trans}}^{-v}(K_-)$, what is

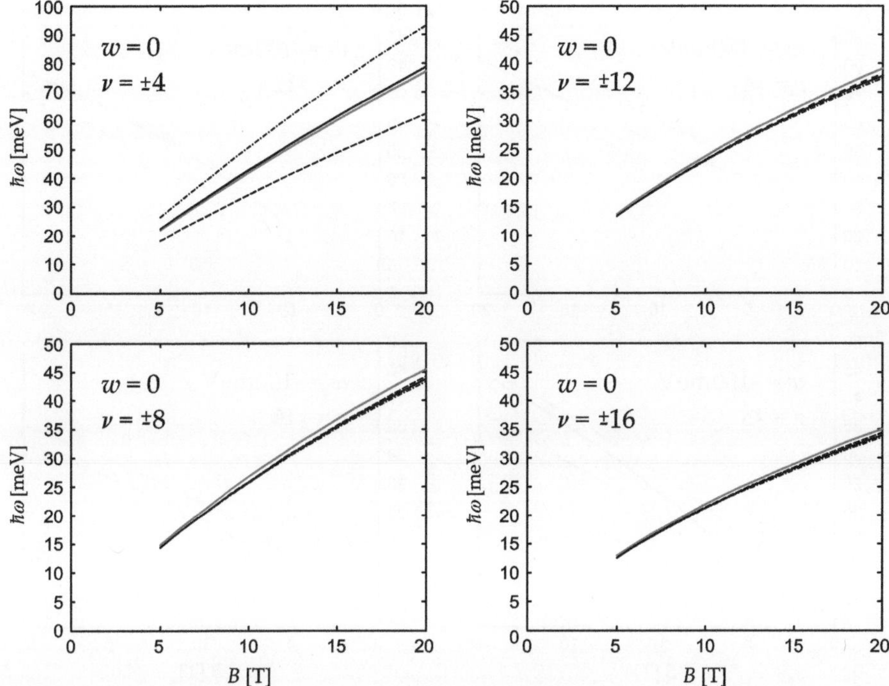

Fig. 4.6 Energy of low-energy inter-LL excitations as a function of magnetic field for $w = 0$. The broken lines are the contributions of individual valleys to the transition energy: *black dashed* and *dot-dashed lines* denote the transition energy for positive (negative) ν at K_+ (K_-) and K_- (K_+), respectively. The *solid black lines* show the contribution of both valleys to the transition energy, calculated according to Eq. (4.27) (in this case the transition energy is the same for both positive and negative ν), whereas *solid grey lines* depict the transition energy in a neutral ($u = 0$) structure. Note that for $\nu = 8, 12, 16$ all *black lines* are very close to each other and difficult to resolve. Figure reprinted from Ref. [29], Copyright (2009), with permission from Elsevier

clearly seen in all four graphs in Fig. 4.6. Moreover, both transitions have the same intensity and contribute equally to $\epsilon^{\nu}_{\text{trans}}$ (shown in black solid line). We can compare our result for $\epsilon^{\nu}_{\text{trans}}$ to that obtained for a neutral bilayer ($u = 0$ at all magnetic fields and filling factors). For the latter case, transition energies are simply given by the spacing between corresponding valley-degenerate Landau levels (see Eq. 4.5) and are plotted in Fig. 4.6 with grey solid lines. We see that non-zero u decreases the energy of the transition. The greater $|u|$ and B, the bigger the difference between the excitation energy in the neutral and biased bilayer graphene. However, this difference decreases with increasing filling factor (this is because the higher energy of the electron states involved, the lesser they are influenced by the perturbation).

Introduction of the parameter w describing initial intrinsic interlayer charge asymmetry breaks symmetry between the conduction and valence band inter-LL excitations as presented in Fig. 4.7 for the case of $w = -100\,\text{meV}$. The valence band excitation has greater energy than the conduction band excitation at filling factor $\nu = \pm4$. However, this situation is reversed for higher filling factors

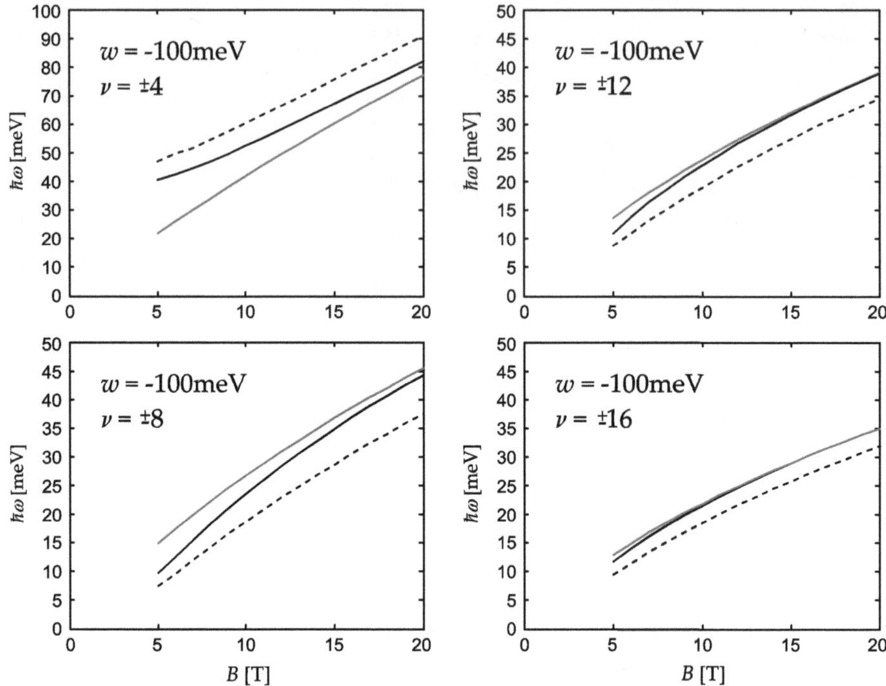

Fig. 4.7 Energy of low-energy inter-LL excitations as a function of magnetic field for $w = -100$ meV. *Black solid* and *dashed lines* denote the transition energy for positive and negative filling factor, respectively. Grey solid lines depict the transition energy in a neutral ($u = 0$) structure. Figure reprinted from Ref. [29], Copyright (2009), with permission from Elsevier

$\nu = \pm 8, \pm 12, \pm 16$ (this reversal was not observed in the experiment [14]). For this specific case, $w = -100$ meV, the asymmetry introduced between excitations for filling factors ν and $-\nu$ is of the size of 3–10 meV.

These two effects: (1) reduction of the transition energy with the increase of u and (2) breaking of the symmetry between transitions for positive and negative filling factor caused by non-zero w, may partly account for the disagreement between experimental findings and transition energies obtained using Eq. (4.5) applicable for neutral bilayers which were used in Ref. [14] to fit the data. Other investigations [38–40] show that additional corrections may arise from electron-electron interactions.

4.3.3 Magneto-Optical Spectra in Charged Bilayer: High-Energy Inter-Landau Level Transitions

Using the self-consistently obtained values of the interlayer asymmetry u for a given filling factor ν and magnetic field B, we can analyse again the optical transition spec-

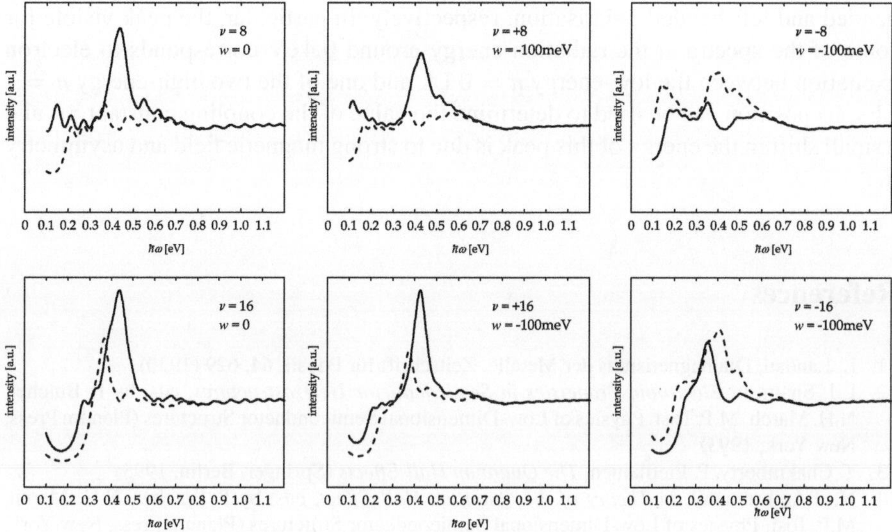

Fig. 4.8 Magneto-optical absorption spectra of bilayer graphene in strong external magnetic field $B = 14$T and for filling factors $\nu = 8$ and $\nu = 16$ (*top* and *bottom* rows, respectively) and the case of $w = 0$ (*left column*) and $w = -100$ meV (*middle* and *right column*). For the symmetric case of $w = 0$, *solid* and *dashed lines* show absorption of right-handed (left-handed) and left-handed (right-handed) circularly polarised light for the positive (negative) filling factor, respectively. For the case of $w = -100$ meV, *solid* and *dashed lines* represent absorption of right and left-handed circularly polarised light, respectively. Figure adapted from Ref. [29], Copyright (2009), with permission from Elsevier

tra corresponding to transitions between LLs in split bands of the bilayers, Sect. 4.2. With the help of Eq. (4.22) describing the intensity of absorption in external magnetic field, we now numerically compute the infrared optical absorption spectra of right ($\oplus$) and left-handed ($\ominus$) circularly polarized light for bilayer graphene in a strong external magnetic field. As opposed to Fig. 4.3c, we describe this time a charged bilayer with significant interlayer asymmetry u. Also, because of $\nu \neq 0$, some of the transitions allowed in the neutral bilayer are blocked because the initial (final) state is empty (filled). The broadening of the Landau levels is again modeled using a Lorentzian shape with the same full width at half maximum $\gamma = 60$ meV for all Landau levels. Numerical results for magnetic field $B = 14$T and filling factors $\nu = 8$ and $\nu = 16$ are shown in Fig. 4.8. For the case of $w = 0$, the symmetry of the system demands that the intensity of absorption of light with a given polarisation for filling factor ν is the same as that of the light with the inverted polarisation for filling factor $-\nu$. This, indeed, is the case for graphs in the left column of Fig. 4.8, where black solid and dashed lines show absorption of right-handed (left-handed) and left-handed (right-handed) circularly polarised light for the positive (negative) filling factor, respectively. Such symmetry is broken for the case of $w = -100$ meV, for which the spectra for positive and negative filling factors are shown in separate panels in Fig. 4.8, where solid and dashed lines refer to right-handed and left-handed polarisation, respectively. In particular, the peak visible for some of the spectra at the radiation energy around $0.4\,\mathrm{eV}$ corresponds to electron excitation between the low-energy $n = 0$ LL and one of the two high-energy $n = 1$ LLs. Its position can be used to determine the value of the coupling constant γ_1, and a small shift in the energy of this peak is due to strong magnetic field and asymmetry u.

References

1. L. Landau, Diamagnetismus der Metalle. Zeitschrift für Physik **64**, 629 (1930)
2. L.J. Sham, in *Electronic Properties in Semiconductor Heterostructures*, eds. by P. Butcher, N.H. March, M.P. Tosi. Physics of Low-Dimensional Semiconductor Structures (Plenum Press, New York, 1993)
3. T. Chakraborty, P. Pietiläinen, *The Quantum Hall Effects* (Springer, Berlin, 1995)
4. N. d'Ambrumenil, in *Theory of the Quantum Hall Effect*, eds. by P. Butcher, N.H. March, M.P. Tosi. Physics of Low-Dimensional Semiconductor Structures (Plenum Press, New York, 1993)
5. K.S. Novoselov, A.K. Geim, S.V. Morozov, D. Jiang, M.I. Katsnelson, I.V. Grigorieva, S.V. Dubonos, A.A. Firsov, Two-dimensional gas of massless dirac fermions in graphene. Nature **438**, 197 (2005)
6. Y. Zhang, Y.W. Tan, H.L. Stormer, P. Kim, Experimental observation of quantum Hall effect and Berry's phase in graphene. Nature **438**, 201 (2005)
7. K.S. Novoselov, E. McCann, S.V. Morozov, V.I. Fal'ko, M.I. Katsnelson, U. Zeitler, D. Jiang, F. Schedin, A.K. Geim, Unconventional quantum Hall effect and Berry's phase of 2pi in bilayer graphene. Nat. Phys. **2**, 177 (2006)
8. Z. Jiang, Y. Zhang, Y.-W. Tan, H.L. Stormer, P. Kim, Quantum Hall effect in graphene. Solid State Commun. **143**, 14 (2007)
9. J.C. Maan, in *Magneto-Optical Properties of Semiconductor Heterostructures*, eds. by P. Butcher, N.H. March, M.P. Tosi. Physics of Low-Dimensional Semiconductor Structures (Plenum Press, New York, 1993)
10. M.L. Sadowski, G. Martinez, M. Potemski, C. Berger, W.A. de Heer, Landau level spectroscopy of ultrathin graphite layers. Phys. Rev. Lett. **97**, 266405 (2006)
11. Z. Jiang, E.A. Henriksen, L.C. Tung, Y.-J. Wang, M.E. Schwartz, M.Y. Han, P. Kim, H.L. Stormer, Infrared spectroscopy of Landau levels of graphene. Phys. Rev. Lett. **98**, 197403 (2007)
12. R.S. Deacon, K.-C. Chuang, R.J. Nicholas, K.S. Novoselov, A.K. Geim, Cyclotron resonance study of the electron and hole velocity in graphene monolayers. Phys. Rev. B **76**, 081406(R) (2007)
13. M.L. Sadowski, G. Martinez, M. Potemski, C. Berger, W.A. de Heer, Magnetospectroscopy of epitaxial few-layer graphene. Solid State Commun. **143**, 123 (2007)
14. E.A. Henriksen, Z. Jiang, L.-C. Tung, M.E. Schwartz, M. Takita, Y.-J. Wang, P. Kim, H.L. Stormer, Cyclotron resonance in bilayer graphene. Phys. Rev. Lett. **100**, 087403 (2008)
15. P. Płochocka, C. Faugeras, M. Orlita, M.L. Sadowski, G. Martinez, M. Potemski, M.O. Goerbig, J.-N. Fuchs, C. Berger, W.A. de Heer, High-energy limit of massless Dirac fermions in multilayer graphene using magneto-optical transmission spectroscopy. Phys. Rev. Lett. **100**, 087401 (2008)
16. M. Orlita, C. Faugeras, P. Plochocka, P. Neugebauer, G. Martinez, D.K. Maude, A.-L. Barra, M. Sprinkle, C. Berger, W.A. de Heer, M. Potemski, Approaching the Dirac point in high mobility multi-layer epitaxial graphene. Phys. Rev. Lett. **101**, 267601 (2008)

17. P. Neugebauer, M. Orlita, C. Faugeras, A.-L. Barra, M. Potemski, How perfect can graphene be? Phys. Rev. Lett. **103**, 136403 (2009)
18. E.A. Henriksen, P. Cadden-Zimansky, Z. Jiang, Z.Q. Li, L.-C. Tung, M.E. Schwartz, M. Takita, Y.-J. Wang, P. Kim, H.L. Stormer, Interaction-induced shift of the cyclotron resonance of graphene using infrared spectroscopy. Phys. Rev. Lett. **104**, 067404 (2010)
19. A.M. Witowski, M. Orlita, R. Stepniewski, A. Wysmolek, J.M. Baranowski, W. Strupinski, C. Faugeras, G. Martinez, M. Potemski, Quasi-classical cyclotron resonance of Dirac fermions in highly doped graphene. Phys. Rev. B **82**, 165305, arXiv:1007.4153 (2010)
20. M. Orlita, M. Potemski, Dirac electronic states in graphene systems: optical spectroscopy studies. Semicond. Sci. Technol. **25**, 063001 (2010)
21. W. Kohn, Cyclotron resonance and de Haas-van Alphen oscillations of an interacting electron gas. Phys. Rev. **123**, 1242 (1961)
22. D.S.L. Abergel, V.I. Fal'ko, Optical and magneto-optical far-infrared properties of bilayer graphene. Phys. Rev. B **75**, 155430 (2007)
23. M. Inoue, Landau levels and cyclotron resonance in graphite. J. Phys. Soc. Jpn **17**, 808 (1962)
24. O.P. Gupta, P.R. Wallace, Effect of trigonal warping on Landau-levels of graphite. Phys. Status Solidi B **54**, 53 (1972)
25. K. Nakao, Landau-level structure and magnetic breakthrough in graphite. J. Phys. Soc. Jpn **40**, 761 (1976)
26. E. McCann, V.I. Fal'ko, Landau level degeneracy and quantum hall effect in a graphite bilayer. Phys. Rev. Lett. **96**, 086805 (2006)
27. D.S.L. Abergel, V. Apalkov, J. Berashevich, K. Ziegler, T. Chakraborty, Properties of graphene: a theoretical perspective. Adv. Phys. **59**, 261 (2010)
28. M. Mucha-Kruczyński, D.S.L. Abergel, E. McCann, V.I. Fal'ko, On spectral properties of bilayer graphene: the effect of an SiC substrate and infrared magneto-spectroscopy. J. Phys. Condens. Matter **21**, 344206 (2009)
29. M. Mucha-Kruczyński, E. McCann, V.I. Fal'ko, Influence of interlayer asymmetry on magneto-spectroscopy of bilayer graphene. Solid State Commun. **149**, 1111 (2009)
30. B.E. Feldman, J. Martin, A. Yacoby, Broken-symmetry states and divergent resistance in suspended bilayer graphene. Nat. Phys. **5**, 889 (2009)
31. E. McCann, Asymmetry gap in the electronic band structure of bilayer graphene. Phys. Rev. B **74**, 161403(R) (2006)
32. E.V. Castro, K.S. Novoselov, S.V. Morozov, N.M.R. Peres, J.M.B. Lopes dos Santos, J. Nilsson, F. Guinea, A. Geim, A.H. Castro Neto, Biased bilayer graphene: semiconductor with a gap tunable by the electric field effect. Phys. Rev. Lett. **99**, 216802 (2007)
33. H. Min, B. Sahu, S.K. Banerjee, A.H. MacDonald, Ab initio theory of gate induced gaps in graphene bilayers. Phys. Rev. B **75**, 155115 (2007)
34. F. Guinea, A.H. Castro Neto, N.M.R. Peres, Electronic states and landau levels in graphene stacks. Phys. Rev. B **73**, 245426 (2006)
35. J.M. Pereira Jr, F.M. Peeters, P. Vasilopoulos, Landau levels and oscillator strength in a biased bilayer of graphene. Phys. Rev. B **76**, 115419 (2007)
36. T. Misumi, K. Shizuya, Electromagnetic response and pseudo-zero-mode landau levels of bilayer graphene in a magnetic field. Phys. Rev. B **77**, 195423 (2008)
37. E.V. Castro, K.S. Novoselov, S.V. Morozov, N.M.R. Peres, J.M.B. Lopes dos Santos, J. Nilsson, F. Guinea, A. Geim, A.H. Castro Neto, Electronic properties of a biased graphene bilayer. J. Phys. Condens. Matter **22**, 175503 (2010)
38. S.V. Kusminskiy, D.K. Campbell, A.H. Castro Neto, Electron-electron interactions in graphene bilayers. Europhys. Lett. **85**, 58005 (2009)
39. D.S.L. Abergel, T. Chakraborty, Long-range Coulomb interaction in bilayer graphene. Phys. Rev. Lett. **102**, 056807 (2009)
40. K. Shizuya, Many-body corrections to cyclotron resonance in monolayer and bilayer graphene. Phys. Rev. B **81**, 075407 (2010)

Chapter 5
Electronic Raman Spectroscopy

As discussed in Sect. 4.3, experimental measurements of the bilayer graphene Landau level structure with infrared absorption showed that tight-binding description for neutral bilayer is unable to describe all the important physics [1]. Some theoretical explanations were suggested, based both on many-body effects and charging effects [2–5], but the issue has not yet been clarified. It would be therefore beneficial to have at one's disposal another probe of the Landau level structure but with different selection rules. Then, electronic excitations between different pairs of levels would be measured. This could help gain more insight into the physics of the problem. In this chapter, we investigate the possibility of using electronic Raman spectroscopy as such a probe. In our analysis, we follow closely the paper of the author, Ref. [6].

The electronic Raman spectroscopy (ERS) can provide information about various single-particle and collective electron excitations in the system under investigation. In semiconductors, it has been, amongst others, employed to investigate donor and acceptor states, plasmons and also spin-density fluctuations involving electron spin-flip due to the spin-orbit interaction [7, 8]. The inelastic scattering of photons on electrons in semiconductor placed in an external magnetic field was first discussed by Wolff, who pointed out that unequal spacing of the Landau levels resulting from nonparabolicity of the electronic bands is crucial for the electron-photon interaction matrix elements not to vanish [9]. The features corresponding to the electronic contribution to the Raman scattering in an external magnetic field were observed in many semiconductors, for example, InSb [10] and GaAs [11]. Very recently, the Raman spectroscopy of electronic excitations in monolayer graphene has been investigated theoretically [12]. It has been shown that at high magnetic fields the inelastic light scattering accompanied by the excitation of the electronic mode with the highest quantum efficiency involves the generation of inter-band electron-hole pairs. At high (quantizing) magnetic fields this leads to the electron excitations from the Landau level n^- (with energy $\epsilon_{n-} = -\sqrt{2n}\,\hbar v/\lambda_B$) in the valence band to the Landau level n^+ ($\epsilon_{n+} = \sqrt{2n}\,\hbar v/\lambda_B$) in the conduction band with energies $\omega_n = 2\sqrt{2n}\,\hbar v/\lambda_B$ and crossed polarisation of in/out photons. This fact should be contrasted with the $\Delta n = \pm 1$ selection rules for transitions between Landau levels in the absorption of

M. Mucha-Kruczyński, *Theory of Bilayer Graphene Spectroscopy*,
Springer Theses, DOI: 10.1007/978-3-642-30936-6_5,
© Springer-Verlag Berlin Heidelberg 2013

left and right-handed circularly polarised infrared photons (first shown in [13], see Sect. 4.2). The need for different selection rules for those two spectroscopies seems natural, if we think in terms of conservation of the orbital momentum in a single scattering process. It is obvious that the necessity to produce an outgoing photon in ERS puts very different constraints on the electronic states that can be involved in the process. Hence, electronic Raman spectroscopy provides data supplementary to that obtained in optical absorption. The question remains whether the signal is strong enough to be experimentally detectable. Again, some insight is given by the results obtained for monolayer graphene in an external magnetic field: it has been shown, among others, that the quantum efficiency I_1 of the lowest peak in the spectrum is $I_1 \sim 10^{-13}$ for $B = 20\,\mathrm{T}$ and for photons with energies in the visible range [12]. Such intensity may be possible to detect in an experiment.

Here, we investigate the contribution of the low-energy electronic excitations towards the Raman spectrum of bilayer graphene for the incoming photon energy $\Omega \gtrsim 1\,\mathrm{eV}$. Starting with the four-band tight-binding model, we derive an effective scattering amplitude that can be easily incorporated into the two-band approximation. We show that due to the influence of the high-energy bands, this effective scattering amplitude is different from the contact interaction amplitude obtained if the two-band model is chosen as the starting point for the theory. We then go on to calculate the spectral density of the inelastic light scattering accompanied by the excitation of electron-hole pairs in bilayer graphene. In the absence of the magnetic field, this contribution is non-zero but constant, reflecting the parabolic dispersion of the low-energy bands in a bilayer crystal. In doped structures a sharp step at twice the Fermi energy is expected as the transitions below that threshold are blocked. In an external magnetic field, similarly to monolayer graphene, the dominant Raman-active modes are the $n^- \to n^+$ inter-Landau-level transitions with crossed polarisation of in/out photons. Finally, we estimate the quantum efficiency of a single $n^- \to n^+$ transition peak in the magnetic field of $10\,\mathrm{T}$ to be $I_{n^- \to n^+} \sim 10^{-12}$ for the incoming photon energy of the order of $1\,\mathrm{eV}$ [6].

5.1 General Considerations

To describe the electronic contribution to Raman scattering, we start by discussing the angle-resolved probability of the Raman scattering, $w(\tilde{q})$. It is given as a ratio of the energy flux of the outgoing photons scattered from the state with the initial momentum q into a state with momentum $\tilde{q}$,[1] $\phi^{\mathrm{out}}_{\tilde{q} \leftarrow q}$, to the energy flux of the incoming photons with the initial momentum q, ϕ^{in}_q,

$$w(\tilde{q}) = \frac{\phi^{\mathrm{out}}_{\tilde{q} \leftarrow q}}{\phi^{\mathrm{in}}_q}. \tag{5.1}$$

---

[1] For further benefit, we treat q and $\tilde{q}$ as two-dimensional and in the plane of the sample. The third, out-of-plane component, only becomes explicitly important in Eq. (5.3).

The energy flux of the incoming photons is given by the number of photons with momentum q, n_q, found in volume V, each carrying energy Ω, $\phi_q^{\text{in}} = \frac{\Omega n_q}{V}$. The energy flux of the outgoing photons is described by the number of photons scattered from q to $\tilde{q}$ in unit time, $\dot{n}_{\tilde{q}}$, and originating from an area S of the material, their energy $\tilde{\Omega}$ and the speed of light, c, as $\phi_{\tilde{q} \leftarrow q}^{\text{out}} = \frac{\tilde{\Omega} \dot{n}_{\tilde{q} \leftarrow q}}{Sc}$. We assume that the only processes creating the outgoing photons in the state with momentum $\tilde{q}$ are those under consideration here and therefore,

$$\dot{n}_{\tilde{q} \leftarrow q} = \frac{S}{2\pi \hbar^3} \int dp \, |\bar{\mathcal{R}}|^2 \, n_q f_i \left(1 - f_f\right) \delta\left(\epsilon_i + \omega - \epsilon_f\right),$$

where i and f denote the initial and final state, respectively, f_i and f_f are filling factors of the corresponding state, $\bar{\mathcal{R}}$ is the scattering amplitude included through the Fermi's golden rule and we integrated out the electron states. Using Eq. 5.1, the angle-resolved probability of the Raman scattering, $w(\tilde{q} \approx \mathbf{0})$, is

$$w = \frac{2V\tilde{\Omega}}{c\pi \hbar^3 \Omega} \int dp |\langle f|\bar{\mathcal{R}}|i\rangle|^2 \times f_i \left(1 - f_f\right) \delta\left(\epsilon_i + \omega - \epsilon_f\right). \tag{5.2}$$

Here, we have already taken into account the spin and valley degeneracies (the latter important in graphene systems).

After adding up contributions from all final states with the same magnitude of the momentum $\tilde{q} = |\tilde{q}|$, we obtain the angle-integrated spectral density of Raman scattering $g(\omega)$,

$$g(\omega) = V \iint \frac{d\tilde{q} d\tilde{q}_z}{(2\pi \hbar)^3} \, w \, \delta\left(\tilde{\Omega} - c\sqrt{\tilde{q}^2 + \tilde{q}_z^2}\right), \tag{5.3}$$

which gives the probability for the incoming photon to scatter inelastically with energy $\tilde{\Omega} = \Omega - \omega$ and describes the shape of the experimental spectrum. Finally, we also consider in this chapter the total quantum efficiency (probability) for an incoming photon to scatter inelastically on an electron with creation of an electron-hole pair, $I = \int d\omega \, g(\omega)$. This last quantity tells us what part of the incoming beam of photons undergoes the inelastic scattering under consideration.

5.2 The Two-Photon Field and the Electron-Photon Interaction

As a starting point for our consideration of ERS for bilayer graphene, we choose the tight-binding Hamiltonian in the linear approximation, Eq. (2.9). We shall keep terms up to quadratic in the electronic momentum p, but disregard the least important couplings, v_4 and γ_n. Then, using the set of Pauli matrices, σ_x, σ_y and σ_z, the Hamiltonian $\hat{H}_\xi(p)$ describing a single electron with momentum p in the vicinity of the valley K_ξ can be written in the form

$$
\hat{H}_\xi(\boldsymbol{p}) = \xi \begin{pmatrix} v_3(\boldsymbol{\sigma}\cdot\boldsymbol{p})^T & v\boldsymbol{\sigma}\cdot\boldsymbol{p} \\ v\boldsymbol{\sigma}\cdot\boldsymbol{p} & \xi\gamma_1\sigma_x \end{pmatrix}
$$
$$
- \mu \begin{pmatrix} \frac{v_3}{v}[\sigma_x(p_x^2 - p_y^2) + 2\sigma_y p_x p_y] & \sigma_x(p_x^2 - p_y^2) - 2\sigma_y p_x p_y \\ \sigma_x(p_x^2 - p_y^2) - 2\sigma_y p_x p_y & 0 \end{pmatrix}. \tag{5.4}
$$

To describe the process of inelastic scattering of light on electrons in our material, we consider an experimental setup in which incoming laser light of energy $\Omega \gg \gamma_1$, in-plane momentum $\boldsymbol{q}$ (out-of-plane component of momentum equal to $q_z = \sqrt{\Omega^2/c^2 - \boldsymbol{q}^2}$) and polarisation $\boldsymbol{l}$ is shined onto to the sample. Scattered photon has polarisation $\tilde{\boldsymbol{l}}$, in-plane momentum $\tilde{\boldsymbol{q}}$ and energy $\tilde{\Omega} = \Omega - \omega$. We also assume the temperature T to be smaller than the Raman shift, $k_B T < \omega$ (k_B is the Boltzmann's constant). As shown in the previous section, all physical quantities of interest are ultimately connected to the quantum mechanical amplitude, $\bar{\mathcal{R}}$, describing the electron transition between the initial and final states. In our case, the inelastic light scattering may occur via (I) a one-step process [so called contact interaction, depicted in Fig. 5.1a] or (II) a two-step process involving an intermediate state [shown in Fig. 5.1b]. The former process is the usual inelastic scattering of an incoming photon on an electron with transfer of energy to the electron. In contrast, the two-step process involves: the absorption (or emission) of a photon with energy Ω ($\tilde{\Omega}$) which transfers an electron with momentum $\boldsymbol{p}$ from an occupied state in the valence band into a virtual intermediate state (energy is not conserved at this stage), followed by emission (or absorption) of the second photon with energy $\tilde{\Omega}$ (Ω) which transfers the electron into the final state. As a result of both a single one or two-step process, an electron-hole pair in the low-energy bands is created with the electron and the hole having almost the same momentum ($\boldsymbol{p} + \boldsymbol{q} - \tilde{\boldsymbol{q}}$ and $\boldsymbol{p}$, respectively), since $\boldsymbol{q}, \tilde{\boldsymbol{q}} \ll \boldsymbol{p}$ and the momentum transfer from light is negligible ($v/c \sim 3\cdot 10^{-3}$). Therefore, $\boldsymbol{p} + \boldsymbol{q} - \tilde{\boldsymbol{q}} \approx \boldsymbol{p}$ and due to the approximately electron-hole symmetric band structure in the vicinity of Brillouin zone corners, the electron initial and final energies ϵ_i and ϵ_f are related, $\epsilon_f \approx -\epsilon_i$.

To include the interaction of the electrons with photons, we construct the canonical momentum $[\boldsymbol{p} - e(\boldsymbol{A}(\boldsymbol{r}, t') + \tilde{\boldsymbol{A}}(\boldsymbol{r}, t''))]$, where $\boldsymbol{A}(\boldsymbol{r}, t')$ and $\tilde{\boldsymbol{A}}(\boldsymbol{r}, t'')$ are the vector potentials of the incoming and outgoing light, respectively,

$$
\boldsymbol{A}(\boldsymbol{r}, t') = \frac{1}{\sqrt{2\epsilon_0 V \Omega}} \left(\boldsymbol{l} e^{i(\boldsymbol{q}\cdot\boldsymbol{r} - \Omega t')/\hbar} b_{\boldsymbol{q},q_z,l} + \text{H.c} \right);
$$

$$
\tilde{\boldsymbol{A}}(\boldsymbol{r}, t'') = \frac{1}{\sqrt{2V\epsilon_0 \tilde{\Omega}}} \left(\tilde{\boldsymbol{l}}^* e^{-i(\tilde{\boldsymbol{q}}\cdot\boldsymbol{r} - \tilde{\Omega} t'')/\hbar} b^\dagger_{\tilde{\boldsymbol{q}},\tilde{q}_z,\tilde{l}} + \text{H.c.} \right),
$$

with $b_{\boldsymbol{q},q_z,l}$ ($b^\dagger_{\boldsymbol{q},q_z,l}$) denoting an annihilation (creation) operator for a photon with in-plane momentum $\boldsymbol{q}$, out-of-plane momentum component q_z and polarisation $\boldsymbol{l}$.

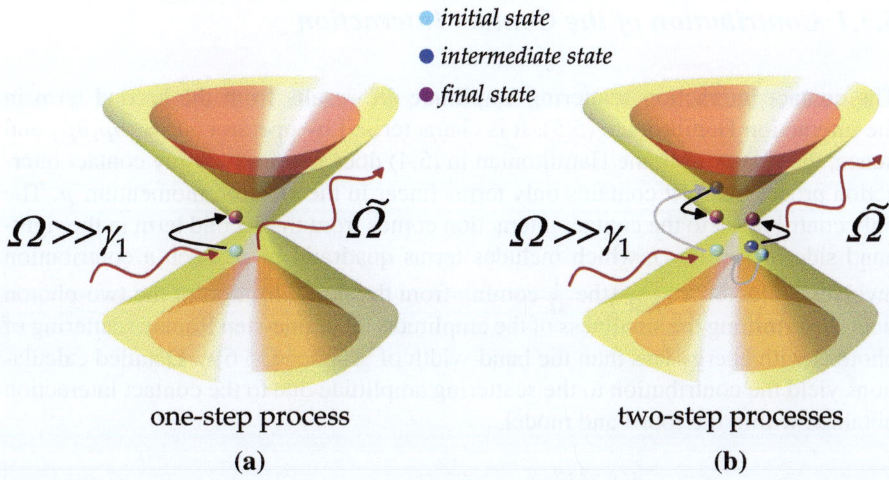

Fig. 5.1 Schematic depiction of the (**a**) one and (**b**) two-step ERS processes considered in this text, shown for the valley K_+. For the two-step processes, *grey* (*black*) *solid lines* indicate the first (second) step of the process. Also, in (**b**), two different cases of the possible two-step process, one involving an intermediate state in the high-energy band (sequence with the *dark blue ball* on the *red* high-energy band) and one only involving states in the low-energy bands, have been shown. The *light blue* (*purple*) *circle* denotes the hole (electron) in the final electron-hole pair, while the *dark blue circle* represents the intermediate virtual state (if relevant). Note that for any intermediate state $|\nu\rangle$ with energy ϵ_ν, Ω, $\tilde{\Omega} \gg \epsilon_\nu$

We then expand the Hamiltonian $\hat{H}_\xi(\boldsymbol{p} - e[\boldsymbol{A} + \tilde{\boldsymbol{A}}])$ up to the second order in the vector potential and write down the interaction part,

$$\hat{H}_{\text{int}} = \hat{\boldsymbol{j}} \cdot \left(\boldsymbol{A}(\boldsymbol{r}, t') + \tilde{\boldsymbol{A}}(\boldsymbol{r}, t'')\right) + \frac{e^2}{2} \sum_{i,j} \frac{\partial^2 \hat{H}_\xi}{\partial p_i \partial p_j} A_i \tilde{A}_j, \tag{5.5}$$

where $\hat{\boldsymbol{j}} = -e \frac{\partial \hat{H}_\xi}{\partial \boldsymbol{p}}$ is the current operator.

5.3 Scattering Amplitude of the ERS Process

We now turn to the calculation of the scattering amplitude $\bar{\mathcal{R}}$. In our case, it is a sum of two contributions: $\delta \mathcal{R}$ due to the one-step processes (contact interaction) and $\mathcal{R}$ due to the two-step processes.

5.3.1 Contribution of the Contact Interaction

The contact interaction scattering amplitude $\delta\mathcal{R}$ results from the second term in the interaction Hamiltonian (5.5). It is characterised by operators $\partial^2 \hat{H}/\partial p_i \partial p_j$ and hence, the first term in the Hamiltonian in (5.4) does not produce any contact interaction processes, as it contains only terms linear in the electron momentum $\boldsymbol{p}$. The only contribution to the contact interaction comes from the second term in the right-hand side of Eq. (5.4), which includes terms quadratic in $\boldsymbol{p}$. Such a contribution involves prefactor $\sim \frac{v^2}{6\gamma_0 \Omega}$ (the $\frac{1}{\Omega}$ coming from the normalisation of the two-photon field) determining the smallness of the amplitude $\delta\mathcal{R}$ of one-step Raman scattering of photons with energy less than the band-width of graphene, $\sim 6\gamma_0$. Detailed calculations yield the contribution to the scattering amplitude due to the contact interaction obtained within the four-band model,

$$\delta\mathcal{R} = \frac{e^2 \hbar^2 v^2}{6\epsilon_0 V \gamma_0 \sqrt{\Omega\tilde{\Omega}}} \mathcal{L}\cdot\boldsymbol{d}; \tag{5.6}$$

$$\boldsymbol{d} = (l_x \tilde{l}_y^* + l_y \tilde{l}_x^*,\ l_x \tilde{l}_x^* - l_y \tilde{l}_y^*); \quad \mathcal{L} = \left(\mathcal{L}_x, \mathcal{L}_y\right);$$

$$\mathcal{L}_x = \begin{pmatrix} -\frac{v_3}{v}\sigma_y & \sigma_y \\ \sigma_y & 0 \end{pmatrix}; \quad \mathcal{L}_y = \begin{pmatrix} -\frac{v_3}{v}\sigma_x & -\sigma_x \\ -\sigma_x & 0 \end{pmatrix}.$$

5.3.2 Contribution of the Two-Step Processes

To find $\mathcal{R}$, we describe a two-step transition which involves an intermediate virtual state $|\nu\rangle$ with energy ϵ_ν, as

$$\mathcal{R} = \frac{-1}{2\epsilon_0 V \sqrt{\Omega\tilde{\Omega}}}$$

$$\times \left\{ \sum_\nu \int_{-\infty}^{\infty}\int_{-\infty}^{t'} e^{\frac{i}{\hbar}(\epsilon_f - \epsilon_\nu)t'} \left(\boldsymbol{j}\cdot\tilde{\boldsymbol{l}}^*\right) e^{\frac{-i}{\hbar}(\tilde{\boldsymbol{q}}\cdot\boldsymbol{r}-\tilde{\Omega}t')} |\nu\rangle\langle\nu| e^{\frac{i}{\hbar}(\boldsymbol{q}\cdot\boldsymbol{r}-\Omega t'')} \left(\boldsymbol{j}\cdot\boldsymbol{l}\right) e^{\frac{i}{\hbar}(\epsilon_\nu - \epsilon_i)t''} dt' dt'' \right.$$

$$\left. + \sum_\nu \int_{-\infty}^{\infty}\int_{-\infty}^{t'} e^{\frac{i}{\hbar}(\epsilon_f - \epsilon_\nu)t'} \left(\boldsymbol{j}\cdot\boldsymbol{l}\right) e^{\frac{i}{\hbar}(\boldsymbol{q}\cdot\boldsymbol{r}-\Omega t')} |\nu\rangle\langle\nu| e^{\frac{-i}{\hbar}(\tilde{\boldsymbol{q}}\cdot\boldsymbol{r}-\tilde{\Omega}t'')} \left(\boldsymbol{j}\cdot\tilde{\boldsymbol{l}}^*\right) e^{\frac{i}{\hbar}(\epsilon_\nu - \epsilon_i)t''} dt' dt'' \right\}. \tag{5.7}$$

We point out that the virtual state $|\nu\rangle$ may belong to any of the four bands, including the high-energy bands. Depending on the accompanying photon process, electron momentum in the state $|\nu\rangle$ is either $\boldsymbol{p} + \boldsymbol{q}$ or $\boldsymbol{p} - \tilde{\boldsymbol{q}}$. In Eq. (5.7), the first (second) term corresponds to processes in which the photon is absorbed (emitted) in the first step and emitted (absorbed) in the second step of the process. Integration over time in those expressions can be performed by changing variables to $\tau = t' - t''$, which varies at the scale of ω^{-1}, $\omega = \Omega - \tilde{\Omega}$, and $\bar{t} = (t' + t'')/2$, which varies at the scale of $\bar{\Omega}^{-1}$, $\bar{\Omega} = (\Omega + \Omega')/2 \gg \omega$. It is important that in the experimentally applicable situation, $\Omega, \tilde{\Omega} \gg \gamma_1$ [14–22]. Also, we concentrate here on the low-energy excitations in the final states with $\omega \ll \gamma_1$ (that is, the initial and final state belong to the low-energy bands). This allows us to expand factors $\frac{1}{\pm\tilde{\Omega}-\epsilon_\nu}$ resulting from the integration over τ in powers of (ϵ_ν / Ω), keeping terms of the order of 1 and (γ_1 / Ω) (the latter appear when the virtual state is taken to be in the high-energy bands) and to perform summation over the intermediate virtual states of the process. Consequently, the amplitude $\mathcal{R}$ takes the form of a matrix

$$\mathcal{R} \approx \frac{e^2 \hbar^2 v^2}{\epsilon_0 V \Omega \sqrt{\Omega\tilde{\Omega}}} \left\{ -i \begin{pmatrix} \sigma_z & 0 \\ 0 & \sigma_z \end{pmatrix} \left(l \times \tilde{l}^* \right)_z + \frac{1}{\Omega} \mathcal{M} \cdot d \right\} \delta\left(\epsilon_f - \epsilon_i - \omega\right), \quad (5.8)$$

$$\mathcal{M} = (\mathcal{M}_x, \mathcal{M}_y),$$

$$\mathcal{M}_x = \begin{pmatrix} \gamma_1 \sigma_y & \xi v \left(\sigma_y p_x + \sigma_x p_y\right) \\ \xi v \left(\sigma_y p_x + \sigma_x p_y\right) & 0 \end{pmatrix},$$

$$\mathcal{M}_y = \begin{pmatrix} \gamma_1 \sigma_x & \xi v \left(\sigma_x p_x - \sigma_y p_y\right) \\ \xi v \left(\sigma_x p_x - \sigma_y p_y\right) & 0 \end{pmatrix}.$$

5.3.3 The Final Form of the Raman Scattering Amplitude

The scattering amplitude $\bar{\mathcal{R}}$ of the Raman process accompanied by electron-hole excitation is a sum of $\mathcal{R}$ and $\delta\mathcal{R}$. From the comparison of the corresponding prefactors $\frac{e^2 \hbar^2 v^2}{6\epsilon_0 \gamma_0 \sqrt{\Omega\tilde{\Omega}}}$, Eq. 5.6, and $\frac{e^2 \hbar^2 v^2}{\epsilon_0 \Omega} \sqrt{\Omega\tilde{\Omega}}$, Eq. 5.8, it follows that $\delta\mathcal{R} \ll \mathcal{R}$, as $6\gamma_0 \gg \Omega$. Hence, we can neglect the contribution of the one-step processes,

$$\bar{\mathcal{R}} = \mathcal{R} + \delta\mathcal{R} \approx \mathcal{R}. \tag{5.9}$$

Also, we are mostly interested in the low-energy properties of our material. To analyse the contribution of electronic modes toward the low-energy part of Raman spectrum with the photon energy shift $\omega < \gamma_1/2$, which is determined by the excitation of the electron-hole pairs in the low-energy bands with $vp \ll \gamma_1$, we use the effective low-energy Hamiltonian, Eq. 2.12 in Sect. 2.3,

$$\hat{H}_{\text{eff}} = -\frac{v^2}{\gamma_1} \left[\left(p_x^2 - p_y^2 \right) \sigma_x + 2 p_x p_y \sigma_y \right]. \tag{5.10}$$

To characterise the excitation of the low-energy modes corresponding to the transitions between low-energy band states described by $\hat{H}_{\text{eff}}$, we take only the part of $\mathcal{R}$ which acts in that two-dimensional Hilbert space, keep terms in the lowest relevant order in $vp/\gamma_1 \ll 1$ and $\gamma_1/\Omega \ll 1$, and write down an effective amplitude $\mathcal{R}_{\text{eff}}$,

$$\mathcal{R}_{\mathrm{eff}} \approx \frac{e^2 \hbar^2 v^2}{\epsilon_0 V \Omega \sqrt{\Omega \tilde{\Omega}}} \left\{ -i\sigma_z \left(l \times \tilde{l}^*\right)_z + \frac{\gamma_1}{\Omega} \left[\sigma_x d_y + \sigma_y d_x\right] \right\}. \qquad (5.11)$$

We point out that the above matrix cannot be obtained within a theory constrained by the two-band approximation from the very beginning. Of course, one can try to define a contact-interaction-like term due to the terms quadratic in the electron momentum p in the Hamiltonian in Eq. (5.10), which will carry a prefactor $\frac{e^2 \hbar^2 v^2}{\epsilon_0 \gamma_1 \Omega}$. Such a prefactor may suggest a greater magnitude of scattering than prefactor $\frac{e^2 \hbar^2 v^2}{\epsilon_0 \Omega^2}$ in the amplitude $\mathcal{R}_{\mathrm{eff}}$ above. However, the scattering amplitude obtained within the two-band model can only be applied to photons with $\Omega < \gamma_1$, as the necessary constraint is that the electron at no stage of the process leaves the energy range where the Hamiltonian (5.10) is applicable, that is energy range $|\epsilon| \ll \gamma_1$. This is a situation not relevant for Raman spectroscopy since it is usually performed with laser beams using $\Omega \sim 1.3$–$2.8\,\mathrm{eV}$ [14–22].

5.4 ERS Spectra in the Absence of the Magnetic Field

With all the important physical quantities introduced in Sect. 5.1, the only remaining step is to use the expression for the effective scattering amplitude $\mathcal{R}_{\mathrm{eff}}$, Eq. (5.11) (the necessary initial and final states can be then expressed like in Eq. (3.16), Sect. 3.3.1), to obtain the following angle-resolved probability of the Raman scattering, $w(\tilde{q} \approx 0)$,[2]

$$w \approx \frac{\gamma_1 e^4 \hbar v^2}{c\epsilon_0^2 V \Omega^4} \left\{ \Xi_s + \frac{\gamma_1^2}{2\Omega^2} \Xi_o \right\} \theta(\omega - 2\mu),$$
$$\Xi_s = \left| l \times \tilde{l}^* \right|^2, \qquad \Xi_o = 1 + \left(l \times l^*\right) \cdot \left(\tilde{l} \times \tilde{l}^*\right). \qquad (5.12)$$

Above, the first term with polarization factor Ξ_s describes the contribution of photons scattered with the same circular polarization as the incoming beam. The second term, with polarization factor Ξ_o, represents the scattered photons with circular polarization opposite to the incoming beam.

In turn, the angle-integrated spectral density of Raman scattering $g(\omega)$ is

$$g(\omega) = 2 \left(\frac{e^2}{4\pi \epsilon_0 \hbar c} \frac{v}{c} \right)^2 \frac{\gamma_1}{\Omega^2} \left\{ 2\Xi_s + \frac{\gamma_1^2}{\Omega^2} \Xi_o \right\} \theta(\omega - 2\mu). \qquad (5.13)$$

---

[2] In the integration over the electronic momentum p we neglected the trigonal warping of the electronic dispersion caused by v_3. As discussed in Sect. 2.3, it is only important for very low energies. The density of states, apart from the vicinity of the Lifshitz transition, remains almost unaffected (see Fig. 2.5).

Constant spectral density g as a function of ω reflects the parabolicity of the low-energy bands and thus, energy-independent density of states in the bilayer. This is different in monolayer graphene, where $g(\omega) \propto \omega$ [12], what reflects the linear dependence of the density of states of electron-hole pairs on energy. The characteristic of monolayer graphene crossed polarisation of in/out photons is retained in the case of the bilayer system.

Experimentally, constant spectral density g in undoped bilayer graphene is impossible to distinguish from a homogeneous background. However, if the chemical potential μ is not at the neutrality point, then transitions with $\omega < 2\mu$ are blocked. Although new processes, resulting in the creation of the intraband electron-hole pair excitations and very small ω, are possible for $\mu \neq 0$, their contribution carries additional prefactor $v/c \sim \frac{1}{300}$ [9] deciding their smallness in comparison to the interband contribution. Explicit calculation performed for the monolayer graphene showed that the quantum efficiency of the intraband transitions was of the order of 10^{-15} [12]. In contrast, for chemical potential $\mu \sim 50\,\mathrm{meV}$ (corresponding to additional carrier density $n_0 \sim 1.5 \times 10^{12}\,\mathrm{cm}^{-2}$), the lost quantum efficiency due to the blocked interband transitions is, according to Eq. (5.13), $\Delta I \sim 10^{-12}$.

5.5 ERS Spectra in Quantizing Magnetic Fields

The quantization of electron states into Landau levels gives the Raman spectrum due to the electronic excitations, a pronounced structure which can be used to detect their contribution experimentally. We only consider here low-energy Landau levels, as at high energies the Landau level broadening due to, for example, electron-phonon interaction, will smear out the LL spectrum. In strong magnetic fields, we are justified to use the low-energy Hamiltonian as shown in Eq. (5.10), with the eigenvalues and eigenstates as discussed in Sect. 4.1.1,

$$\epsilon_{n\alpha} = \alpha \frac{2\hbar^2 v^2}{\gamma_1 \lambda_B^2} \sqrt{n(n-1)},$$

$$\Psi_{n\alpha} = \begin{pmatrix} \psi_n \\ 0 \end{pmatrix} \text{ for } n = 0, 1, \quad \Psi_{n\alpha} = \frac{1}{\sqrt{2}} \begin{pmatrix} \psi_n \\ \alpha\psi_{n-2} \end{pmatrix} \text{ for } n \geq 2, \qquad (5.14)$$

where $\lambda_B = \sqrt{\hbar/eB}$ is the magnetic length, n is the Landau level index, $\alpha = \pm$ denotes the conduction $(+)$ or valence $(-)$ band. Also, ψ_n is the normalised n-th Landau level wavefunction. We recall that in a neutral bilayer, all LLs have additional four-fold degeneracy (two due to the electron spin and two due to the valley). Moreover, levels $n = 0$ and $n = 1$ are degenerate at $\epsilon = 0$ giving rise to an 8-fold degenerate LL.

We can project our effective transition amplitude $\mathcal{R}_{\mathrm{eff}}$ onto the eigenstates $\Psi_{n\alpha}$ to find the electronic Raman spectrum in the presence of a strong external magnetic field. This leads to the following selection rules for allowed electronic transitions

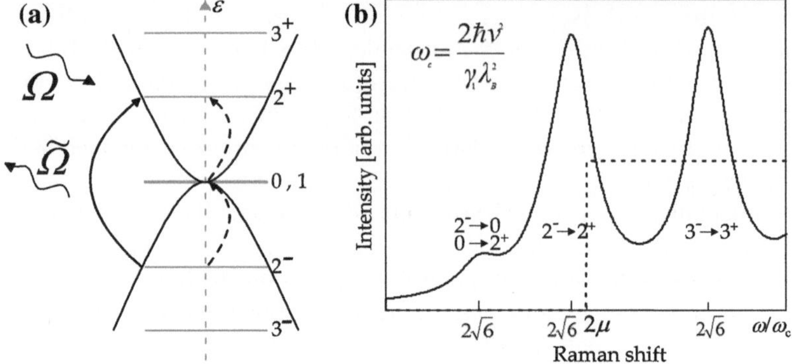

Fig. 5.2 **a** Schematic drawing of allowed inter-LL transitions accompanying the Raman scattering. The *solid* (*dashed*) *line* represents the first dominant (weaker) transition $2^- \leftarrow 2^+$ (pair $2^- \leftarrow 0$ and $0 \leftarrow 2^+$). **b** The low-energy electronic contribution to the Raman spectrum in bilayer graphene. The *solid* (*dashed*) *line* represents the spectrum in the presence (absence) of an external magnetic field and chemical potential $\mu = 0$ ($\mu \neq 0$). For the spectrum in a magnetic field, corresponding inter-LL transitions have been attributed to each peak. Figure reprinted from Ref. [6], Copyright (2010), with permission from APS

from the initial level n^-:

$$(i)\, n^- \to n^+; \qquad (ii)\, (n \mp 1)^- \to (n \pm 1)^+. \tag{5.15}$$

Among those, (i) is the dominant transition. These selection rules, represented schematically in Fig. 5.2, confirm our expectations and show explicitly that using Raman spectroscopy, one can probe different electronic excitations than in optical spectroscopy, where the selection rules are $\Delta n = \pm 1$ ([13, 23], see Sect. 4.2). For a neutral bilayer, the angle-integrated spectral density $g(\omega)$ of Raman scattering in the magnetic field is equal to

$$g(\omega) \approx 16\, \Xi_s \left(\frac{e^2}{4\pi \epsilon_0 \hbar c} \frac{v}{c} \right)^2 \left(\frac{\hbar v}{\lambda_B \Omega} \right)^2 \sum_{n \geq 2} \gamma(\omega - 2\epsilon_{n^+}) + \delta g(\omega),$$

$$\delta g(\omega) = 8\, \Xi_o \left(\frac{\gamma_1}{\Omega} \right)^2 \left(\frac{e^2}{4\pi \epsilon_0 \hbar c} \frac{v}{c} \right)^2 \left(\frac{\hbar v}{\lambda_B \Omega} \right)^2$$

$$\left[\sum_{n=1,2} 2\gamma(\omega - \epsilon_{(n+1)^+}) + \sum_{n \geq 3} \gamma(\omega - \epsilon_{(n+1)^+} - \epsilon_{(n-1)^+}) \right]. \tag{5.16}$$

Here, we used Lorentzian $\gamma(x) = \pi^{-1}\Gamma/(x^2 + \Gamma^2)$ with a width specified by Γ to model the broadening of Landau levels. The term $\delta g(\omega)$ describes the spectral density of the $(n \mp 1)^- \to (n \pm 1)^+$ transitions, which is a correction to the dominant

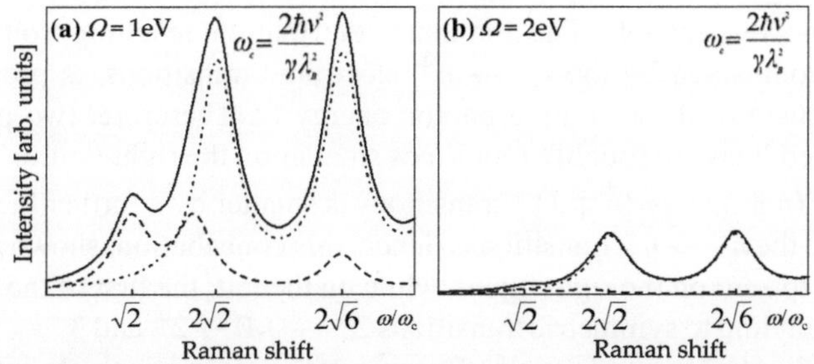

Fig. 5.3 Comparison of electronic contributions to the Raman spectra in neutral bilayer graphene for two different energies of incoming photons: (**a**) $\Omega = 1\,eV$, (**b**) $\Omega = 2\,eV$. For each case, total spectral density $g(\omega)$ and contributions due to the $n^- \to n^+$ and $(n \pm 1)^- \to (n \mp 1)^+$ modes are shown in the *solid*, *dot-dashed* and *dashed line*, respectively. Intensity scale is the same on (**a**) and (**b**); values of the parameters used: $v = 10^6$ m/s, $\gamma_1 = 0.4\,eV$, $B = 10\,T$, and $\Gamma = 0.012\,eV$. Figure reprinted from Ref. [6], Copyright (2010), with permission from APS

contribution due to the $n^- \to n^+$ transitions given by the first term on the right hand side of Eq. (5.16).

An example of the low-energy electronic contribution to the Raman spectrum in the neutral bilayer in strong magnetic field is shown with a solid line in Fig. 5.2b. The dominant features are peaks due to the $n^- \to n^+$ transitions with the first being the $2^- \to 2^+$ transition. We remind again that within the LL indexing scheme applied here, indices 0 and 1 are only used to denote one valley-degenerate level each and no α index is needed for them. The quantum efficiency of a single $n^- \to n^+$ peak in Fig. 5.2b is approximately

$$I_{n^- \to n^+} \approx \left(\frac{v^2}{c^2} \frac{e^2/\lambda_B}{\epsilon_0 \pi \Omega} \right)^2 = \frac{v^4 e^5 B}{\pi^2 c^4 \epsilon_0^2 \hbar \Omega^2} \tag{5.17}$$

per incoming photon, which at the field $B \sim 10\,T$ gives $I_{n^- \to n^+} \sim 10^{-12}$ for $\Omega \sim 1\,eV$ photons, comparable to similar transitions in monolayer graphene [12].

A weaker feature in Fig. 5.2b is the first and the only visible $(n \mp 1)^- \to (n \pm 1)^+$ peak due to both $2^- \to 0$ and $0 \to 2^+$ transitions, positioned to the left of the $2^- \to 2^+$ peak. The quantum efficiencies of the $(n \pm 1)^- \to (n \mp 1)^+$ transitions are smaller by the factor $\left(\frac{\gamma_1}{\Omega} \right)^2$ in comparison to the $n^- \to n^+$ transitions. This is different from the monolayer graphene case, where the corresponding ratio between quantum efficiencies of $(n \pm 1)^- \to (n \mp 1)^+$ and $n^- \to n^+$ transitions is $\left(\frac{\omega}{\Omega} \right)^2$, much smaller than for the bilayer. The term $\delta g(\omega)$ can be further emphasized by changing the energy of incoming photons Ω. Shown in Fig. 5.3a, b, is a comparison of the total spectral density $g(\omega)$ and contributions due to each mode separately, for two different energies of incoming photons, $\Omega = 2\,eV$ and $\Omega = 1\,eV$. The intensity scale is the same on both figures and in each case, the total spectral density $g(\omega)$, the contributions due to the $n^- \to n^+$ and $(n \pm 1)^- \to (n \mp 1)^+$ modes are shown in the solid, dot-dashed and dashed line, respectively. The dominant contribution, resulting from the Raman scattering accompanied by the $n^- \rightarrow n^+$ electronic transitions, is proportional to the inverse square of the incoming photon energy Ω. Therefore, two peaks drawn with dot-dashed lines are roughly four times smaller on the right figure. The spectral density of the $(n \pm 1)^- \rightarrow (n \mp 1)^+$ transitions is smaller by a further factor $\left(\frac{\gamma_1}{\Omega}\right)^2$ in comparison to the $n^- \rightarrow n^+$ transitions. Hence, this contribution, shown with dashed lines, is close to zero on the right figure, while on the left, the first of the two smaller peaks corresponding to symmetric transitions $2^- \rightarrow 0, 0 \rightarrow 2^+$ and $3^- \rightarrow 1, 1 \rightarrow 3^+$ is still visible in the total spectral density. Because of the contrasting polarization factors in Eq. (5.16), contributions of different modes, $n^- \rightarrow n^+$ or $(n \pm 1)^- \rightarrow (n \mp 1)^+$, to the total spectral density could be separated using polarizers. If the polarizers were set as to collect only photons with circular polarization identical to that of the incoming photons, then the $n^- \rightarrow n^+$ contribution would be measured. However, if only the photons with polarization opposite to the polarization of the incoming beam were detected, the $(n \mp 1)^- \rightarrow (n \pm 1)^+$ contribution would be determined.

Increasing the filling factor leads first to the $2^- \rightarrow 0$ and $3^- \rightarrow 1$ transitions being blocked when LLs with $n = 0$ and $n = 1$ are completely filled. Therefore, the height of the two corresponding $(n \pm 1)^- \rightarrow (n \mp 1)^+$ peaks is halved (transitions $0 \rightarrow 2^+$ and $1 \rightarrow 3^+$ are still allowed). Next to disappear are the first $n^- \rightarrow n^+$ peak, that is $2^- \rightarrow 2^+$, and the remains of the first $(n \pm 1)^- \rightarrow (n \mp 1)^+$ peak, (due to the $0 \rightarrow 2^+$ transition) because of the filled LL 2^+. Complete filling of each following Landau level results in the disappearance of the next $n^- \rightarrow n^+$ and $(n \pm 1)^- \rightarrow (n \mp 1)^+$ peaks.

References

1. E.A. Henriksen, Z. Jiang, L.-C. Tung, M.E. Schwartz, M. Takita, Y.-J. Wang, P. Kim, H.L. Stormer, Cyclotron resonance in bilayer graphene. Phys. Rev. Lett. **100**, 087403 (2008)
2. D.S.L. Abergel, T. Chakraborty, Long-range Coulomb interaction in bilayer graphene. Phys. Rev. Lett. **102**, 056807 (2009)
3. S.V. Kusminskiy, D.K. Campbell, A.H. Castro Neto, Electron-electron interactions in graphene bilayers. Europhys. Lett. **85**, 58005 (2009)
4. K. Shizuya, Many-body corrections to cyclotron resonance in monolayer and bilayer graphene. Phys. Rev. B **81**, 075407 (2010)
5. M. Mucha-Kruczyński, E. McCann, V.I. Fal'ko, Influence of interlayer asymmetry on magneto-spectroscopy of bilayer graphene. Solid State Commun. **149**, 1111 (2009)
6. M. Mucha-Kruczyński, O. Kashuba, V.I. Fal'ko, Spectral features due to inter-Landau-level transitions in the Raman spectrum of bilayer graphene, Phys. Rev. B, **82**, 045405 (2010)
7. G. Abstreiter, M. Cardona, A. Pinczuk, in *Light Scattering by Free Carrier Excitations in Semiconductors*, ed. by M. Cardona, G.Güntherodt. Light Scattering in Solids, vol IV (Springer, Heidelberg, 1984)
8. M.V. Klein, in *Electronic Raman Scattering*, ed. by M. Cardona. Light Scattering in Solids, vol I (Springer, Heidelberg, 1983)
9. P.A. Wolff, Thomson and Raman scattering by mobile electrons in crystals. Phys. Rev. Lett. **16**, 225 (1966)
10. R.E. Slusher, C.K.N. Patel, P.A. Fleury, Inelastic light scattering from Landau-level electrons in semiconductors. Phys. Rev. Lett. **18**, 77 (1967)

11. C.K.N. Patel, R.E. Slusher, Light scattering from electron plasmas in a magnetic field. Phys. Rev. Lett. **21**, 1563 (1968)
12. O. Kashuba, V.I. Fal'ko, Signature of electronic excitations in the Raman spectrum of graphene, Phys. Rev. B **80**, 241404(R) (2009)
13. D.S.L. Abergel, V.I. Fal'ko, Optical and magneto-optical far-infrared properties of bilayer graphene. Phys. Rev. B **75**, 155430 (2007)
14. A.C. Ferrari, J.C. Meyer, V. Scardaci, C. Casiraghi, M. Lazzeri, F. Mauri, S. Piscanec, D. Jiang, K.S. Novoselov, S. Roth, A.K. Geim, Raman spectrum of graphene and graphene layers. Phys. Rev. Lett. **97**, 187401 (2006)
15. A. Gupta, G. Chen, P. Joshi, S. Tadigadapa, P.C. Eklund, Raman scattering from high-frequency phonons in supported n-graphene layer films. Nano Lett. **6**, 2667 (2006)
16. D. Graf, F. Molitor, K. Ensslin, C. Stampfer, A. Jungen, C. Hierold, L. Wirtz, Spatially resolved Raman spectroscopy of single- and few-layer graphene. Nano Lett. **7**, 238 (2007)
17. L.M. Malard, J. Nilsson, D.C. Elias, J.C. Brant, F. Plentz, E.S. Alves, A.H. Castro Neto, M.A. Pimenta, Probing the electronic structure of bilayer graphene by Raman scattering, Phys. Rev. B **76**, 201401(R) (2007)
18. J. Yan, E.A. Henriksen, P. Kim, A. Pinczuk, Observation of anomalous phonon softening in bilayer graphene. Phys. Rev. Lett. **101**, 136804 (2008)
19. L.M. Malard, D.C. Elias, E.S. Alves, M.A. Pimenta, Observation of distinct electron-phonon couplings in gated bilayer graphene. Phys. Rev. Lett. **101**, 257401 (2008)
20. A. Das, B. Chakraborty, S. Piscanec, S. Pisana, A.K. Sood, A.C. Ferrari, Phonon renormalization in doped bilayer graphene. Phys. Rev. B **79**, 155417 (2009)
21. Z. Ni, L. Liu, Y. Wang, Z. Zheng, L.-J. Li, T. Yu, Z. Shen, G-band Raman double resonance in twisted bilayer graphene: evidence of band splitting and folding. Phys. Rev. B **80**, 125404 (2009)
22. D.L. Mafra, L.M. Malard, S.K. Doorn, H. Htoon, J. Nilsson, A.H. Castro Neto, M.A. Pimenta, Observation of the Kohn anomaly near the K point of bilayer graphene, Phys. Rev. B **80**, 241414(R) (2009)
23. M. Mucha-Kruczyński, D.S.L. Abergel, E. McCann, V.I. Fal'ko, On spectral properties of bilayer graphene: the effect of an SiC substrate and infrared magneto-spectroscopy. J. Phys. Condens. Matter **21**, 344206 (2009)

Chapter 6
Conclusions

In the previous chapters, we developed theories based on the tight-binding model for π electrons, in order to describe the angle-resolved photoemission, magneto-optical absorption and electronic Raman spectra of bilayer graphene. The results for the first two can be compared to experimental spectra. Specifically, the constant-energy ARPES maps found in literature (in particular in the supporting on-line material to the article by Ohta and co-workers [1]), closely resemble those shown in the Figs. 3.3, 3.4 and 3.6. First of all, we can test our conclusion about the importance of the electronic states in the high-energy bands for the intensity of bilayer graphene ARPES maps. Indeed, asymmetric intensity of the two peaks is always observed in experiment. The change in the density of states due to symmetry breaking parameters, in particular the interlayer asymmetry u, has also been observed with ARPES. The constant-energy maps are then modified very much like shown by our model. The experimentally obtained ARPES spectra were also detailed enough for the magnitude of the direct interlayer coupling γ_1 to be extracted. Unfortunately, little experimental data presented in the aforementioned works do not allow testing of other predictions of our model, that is, whether the sign of γ_1 as well as magnitude and sign of γ_3 can be determined.

In the case of the magneto-optical absorption, the spectra shown in Figs. 4.3c and 4.8 can be compared to those predicted and experimentally measured for optical absorption [2–6]. The main feature—the peak corresponding to the onset of the transitions between the low-energy and split bands—is similar in both cases and allows for an independent check on the value of the coupling γ_1. However, it is very difficult to comment on the relevance of the main point of the considerations in Chap. 4—the contribution of the gate-induced interlayer asymmetry to the interpretation of the experimental results of the work of Henriksen et al. [7]. Although the presented theory suggests the presence of the gap, it is definitely not the only factor responsible for the observed deviations from the 'neutral bilayer' model. In fact, some of the many-body theories show significantly better agreement with the experiment [8, 9].

The situation is different for the electronic Raman spectroscopy. Although inelastic scattering of light is a widely used experimental technique for characterisa-

M. Mucha-Kruczyński, *Theory of Bilayer Graphene Spectroscopy*,
Springer Theses, DOI: 10.1007/978-3-642-30936-6_6,
© Springer-Verlag Berlin Heidelberg 2013

tion of carbon materials and has been employed for graphene systems specifically to investigate not only the electron-phonon coupling but also number of layers, disorder or doping level, the electronic contribution to the Raman spectra has not yet been examined in detail. Therefore, theory presented in Chap. 5 goes one step beyond being tested by comparison to available measurements. The absolute numbers predicted for the quantum efficiencies of the inelastic light scattering accompanied by electron-hole excitations suggest that some of the features discussed may be observable. In our opinion, in the light of the complications in the interpretation of the experimental results of experiments of the type of the one by Henriksen and co-workers, examination of the ERS spectra in an external magnetic field would provide an important additional way of investigating the Landau level structure of bilayer graphene. Departures from the theory presented here could then strengthen the initial claim made in Ref. [7] about the influence of the many-body physics on the inter-LL transitions. This would also give more insight on the usefulness and limits of the tight-binding approach for graphene systems.

Note, that in all three cases presented, it was important to choose as the starting point for the theory, the four-band model. The two-band approximation, neglecting the influence of the high-energy states, does not capture essential features of the spectra discussed in any of the chapters. For ARPES, it cannot describe the asymmetry in the intensity between the two-peak pattern (see Fig. 3.3 and discussion in Sect. 3.3.2), which persists to energies much smaller than γ_1. The high-energy bands are essential in order to obtain the self-consistent values of the interlayer asymmetry both in the absence and presence of the magnetic field. The two-band approximation, for obvious reasons, also cannot describe the inter-Landau-level transitions between the low-energy and split bands caused by absorption of incoming photons. Finally, for inelastic scattering of photons, we showed that the ERS scattering amplitude for experimentally relevant energies of the incoming beam cannot be properly described within the two-band approximation alone. Due to the importance of the high-energy bands in the two-step processes, proper results can only be obtained if the full four-band model is used as the starting point. All those arguments show how carefully any problem related to the electronic structure of bilayer graphene has to be considered before only the two-band approximation is used to predict or explain its results. Good understanding of the physical processes involved is required if the theoretical model is to be correct and useful.

References

1. T. Ohta, A. Bostwick, T. Seyller, K. Horn, E. Rotenberg, Controlling the electronic structure of bilayer graphene. Science **313**, 951 (2006)
2. D.S.L. Abergel, V.I. Fal'ko, Optical and magneto-optical far-infrared properties of bilayer graphene. Phys. Rev. B **75**, 155430 (2007)
3. E.J. Nicol, J.P. Carbotte, Optical conductivity of bilayer graphene with and without an asymmetry gap. Phys. Rev. B **77**, 155409 (2008)

4. L.M. Zhang, Z.Q. Li, D.N. Basov, M.M. Fogler, Z. Hao, M.C. Martin, Determination of the electronic structure of bilayer graphene from infrared spectroscopy. Phys. Rev. B **78**, 235408 (2008)
5. Z.Q. Li, E.A. Henriksen, Z. Jiang, Z. Hao, M.C. Martin, P. Kim, H.L. Stormer, D.N. Basov, Band structure asymmetry of bilayer graphene revealed by infrared spectroscopy. Phys. Rev. Lett. **102**, 037403 (2009)
6. K.F. Mak, C.H. Lui, J. Shan, T.F. Heinz, Observation of an electric-field-induced band gap in bilayer graphene by infrared spectroscopy. Phys. Rev. Lett. **102**, 256405 (2009)
7. E.A. Henriksen, Z. Jiang, L.-C. Tung, M.E. Schwartz, M. Takita, Y.-J. Wang, P. Kim, H.L. Stormer, Cyclotron resonance in bilayer graphene. Phys. Rev. Lett. **100**, 087403 (2008)
8. D.S.L. Abergel, T. Chakraborty, Long-range Coulomb interaction in bilayer graphene. Phys. Rev. Lett. **102**, 056807 (2009)
9. K. Shizuya, Many-body corrections to cyclotron resonance in monolayer and bilayer graphene. Phys. Rev. B **81**, 075407 (2010)

Curriculum Vitae

Marcin Mucha-Kruczyński

A.1 Education

07/2007–01/2011	**Ph.D. in Theoretical Physics**, *Lancaster University*, Lancaster, UK.
10/2002–05/2007	**M.Sc. in Material Engineering**, *Adam Mickiewicz University*, Poznań, Poland.
10/2005–06/2006	Socrates/Erasmus exchange student, *Lancaster University*, Lancaster, UK.

A.2 Experience

Research

10/2010–onwards	**Research Associate**, *Physics Department, Lancaster University*, Lancaster, UK.
10/2010–09/2011	**EPSRC Ph.D. + Fellow**, *Physics Department, Lancaster University*, Lancaster, UK.

Teaching

10/2011–onwards	**Undergraduate Tutor**, *Physics Department, Lancaster University*, Lancaster, UK.
10/2007–06/2010	**Postgraduate Teaching Assistant**, *Physics Department, Lancaster University*, Lancaster, UK.

M. Mucha-Kruczyński, *Theory of Bilayer Graphene Spectroscopy*,
Springer Theses, DOI: 10.1007/978-3-642-30936-6,
© Springer-Verlag Berlin Heidelberg 2013

A.3 Publications

[8] A. S. Mayorov, D. C. Elias, **M. Mucha-Kruczyński**, R. V. Gorbachev, T. Tudorovskiy, A. Zhukov, S. V. Morozov, V. I. Fal'ko, M. I. Katsnelson, A. K. Geim, K. S. Novoselov, *Interaction-Driven Spectrum Reconstruction in Bilayer Graphene*, Science **333**, 860 (2011).

[7] **M. Mucha-Kruczyński**, I. L. Aleiner, and V. I. Fal'ko, *Landau levels in deformed bilayer graphene at low magnetic fields*, Sol. St. Commun. **151**, 1088 (2011).

[6] **M. Mucha-Kruczyński**, I. L. Aleiner, and V. I. Fal'ko, *Strained bilayer graphene: Band structure topology and Landau level spectrum*, Phys. Rev. B **84**, 041404 (2011).

[5] **M. Mucha-Kruczyński**, O. Kashuba, and V. I. Fal'ko, *Spectral features due to inter- Landau-level transitions in the Raman spectrum of bilayer graphene*, Phys. Rev. B **82**, 045405 (2010).

[4] **M. Mucha-Kruczyński**, E. McCann, and V. I. Fal'ko, *Electron-hole asymmetry and energy gaps in bilayer graphene*, Semicond. Sci. Technol. **25**, 033001 (2010).

[3] **M. Mucha-Kruczyński**, E. McCann, and V. I. Fal'ko, *Influence of interlayer asymmetry on magneto-spectroscopy of bilayer graphene*, Sol. St. Commun. **149**, 1111 (2009).

[2] **M. Mucha-Kruczyński**, D. S. L. Abergel, E. McCann, and V. I. Fal'ko, *On spectral properties of bilayer graphene: the effect of an SiC substrate and infrared magnetospectroscopy*, J. Phys.: Condens. Matt. **21**, 344206 (2009).

[1] **M. Mucha-Kruczyński**, O. Tsyplyatyev, A. Grishin, E. McCann, V. I. Fal'ko, A. Bostwick, and E. Rotenberg, *Characterization of graphene through anisotropy of constantenergy maps in angle-resolved photoemission*, Phys. Rev. B **77**, 195403 (2008).

A.4 Invited Talks

- *Spectrum Reconstruction in Bilayer Graphene (due to strain and/or electron–electron interaction)* at the American Physical Society March Meeting 2012 (Boston, US, 27/02– 02/03/2012).
- *Low-energy electronic dispersion in bilayer graphene under uniaxial strain* at the workshop "Quantum phenomena in graphene, other low-dimensional materials, and optical lattices" (Erice, Sicily, 04–07/07/2011).
- *Theory of the Angle-Resolved Photoemission Spectroscopy of Bilayer Graphene* at the NoWNANO Summer School 2010 (Bolton, UK, 24–25/06/2010).

A.5 Other Oral Presentations

- Talk during the Physics of Graphene program at the Kavli Institute for Theoretical Physics, UCSB (Santa Barbara, US, 12/03/2012).
- UK–Japan Workshop and Graphene Roadmap Consultation: Graphene Synthesis and Characterisation for applications (Windermere, UK, 15–18/11/2011).
- Graphene Week 2011 (Obergurgl, Austria, 24–29/04/2011).
- UK–Japan Graphene Workshop (Lancaster, UK, 03–04/02/2011).
- AGRESSIA-09: Assembly on Graphene Electronic Structure—Science & Innovation Award (Keswick, UK, 09–11/10/2009).
- Final ESF-FoNE Conference (Miraflores de la Sierra, Madrid, Spain, 09–13/09/2009).

A.6 Awards and Honours

- Ph.D. thesis selected for publication in the *Springer Theses* series.
- EPSRC Ph.D. + Fellowship (October 2010).
- Chosen for poster presentation at the SET for Britain 2010 exhibition for early-stage and early-career research scientists, engineers and technologists in the House of Commons in London, UK, 08/03/2010.
- Student Poster Prize at IOP Theory of Condensed Matter Group Meeting 2007, Warwick, UK, 17/12/2007.
- United Kingdom's Engineering and Physical Sciences Research Council (EPSRC) scholarship for Ph.D. study.
- Polish Minister of Education and Sport scholarship for outstanding academic achievements (academic year 2005/06).
- Adam Mickiewicz University Faculty of Physics scientific scholarship (academic years 2003/04; 2004/05; 2006/07).

A.7 Referee of the Following Journals

Nanotechnology.
Journal of Physics D: Applied Physics.
New Journal of Physics.
Journal of Vacuum Science and Technology B.

A.8 Memberships

2008–2011   Associate Member of Institute of Physics.

A.9 Languages

Polish   Fluent in speech and writing.
English   Fluent in speech and writing.
German   Elementary.

If you have any concerns about our products,
you can contact us on
ProductSafety@springernature.com

In case Publisher is established outside the EU,
the EU authorized representative is:
Springer Nature Customer Service Center GmbH
Europaplatz 3, 69115 Heidelberg, Germany

Printed by Libri Plureos GmbH
in Hamburg, Germany